To Prof. D. J. Nolte'

With the author's affectionate regards
and sincere thanks for starting off
an interest in Acridation so many years
ago

As ever
[illegible]

BULLETIN OF
THE BRITISH MUSEUM (NATURAL HISTORY)
ZOOLOGY Supplement 6

ERRATA

Page 17, fifth paragraph, second to last line:
For *Saroherodon* *read* *Sarotherodon*.

Page 27, third line from top:
For antividuals *read* individuals.

fourth line from top:
For inderior *read* anterior.

Page 68, second line from top:
For definition *read* dentition

Cichlid fishes of Lake Victoria, East Africa

Cichlid fishes of Lake Victoria, East Africa: the biology and evolution of a species flock

P. H. Greenwood, D.Sc.
Department of Zoology
British Museum (Natural History)

Trustees of the British Museum (Natural History)
London 1974

This volume is reprinted from the *Bulletin of the British Museum (Natural History)*, Zoology series, Supplement 6

ISBN 0 565 00761 0

Printed in Great Britain by John Wright and Sons Ltd. at The Stonebridge Press, Bristol BS4 5NU

Preface

The majority of the papers published in the *Bulletin of the British Museum (Natural History)*, Zoology series and its supplements are essentially taxonomic and therefore of interest to specialists in particular groups. But the results of taxonomy frequently lead to major reappraisals in other areas of biology, and some numbers of the *Bulletin*, and particularly its supplements, are concerned with topics of more general interest to natural historians and evolutionary biologists. For this reason, the present work, published as Supplement 6, is issued simultaneously in a more permanent bound form.

This volume by Dr P. H. Greenwood is concerned with explosive speciation and adaptive radiation. It is an important contribution to the evolutionary question: Why have some plants and animals shown a sudden explosive radiation of forms, while others have plodded conservatively on? There is no dearth of concepts to account for this, but the taxonomy – and the biological and ecological data which this unlocks – must provide the essential basis if theories are to have any meaning.

For some twenty years Dr Greenwood has been painstakingly unravelling the taxonomy of the Lake Victoria Cichlidae – a battle of attrition almost – but the results provide an insight into the way that a lake of less than a million years' standing can bring forth a flock of 150–170 species of cichlid fishes, of which all but eight belong to the genus *Haplochromis*. The taxonomic, biological and ecological data are synthesized to provide the recurrent theme: adaptive radiation in feeding habits, here used as the principal tool in constructing the phylogeny of the group. Many of the conclusions will find a place outside ichthyology if for no other reason that such complex species flocks have rarely been studied in such a penetrating manner.

Finally, it must be noted that the author makes no claim to the work's completeness. Indeed it is his hope that this study will stimulate and guide further research into the cichlid species flocks of African lakes.

J. G. Sheals
Keeper of Zoology

for M.G.

Contents

The Cichlid fishes of Lake Victoria, East Africa: the biology and evolution of a species flock

SYNOPSIS

The fish fauna of Lake Victoria is dominated, both ecologically and in species numbers, by over a hundred endemic species of the cichlid genus *Haplochromis*. This *Haplochromis* species flock has evolved within the developing lake basin during the last three-quarter million years, and its member species now occupy virtually all ecological and trophic niches within the lake, sometimes sharing these niches with species of other families and always sharing them with other and related *Haplochromis* species. The ecological and evolutionary implications of this situation are considered.

The limnological and historical backgrounds to this explosive speciation are reviewed, as are the feeding and breeding habits of the *Haplochromis* species. Morphological adaptations are described and considered in relation to the speed with which the flock evolved. Hypotheses relating to this aspect of the problem and to the phylogeny of the species are developed and, where possible, reviewed in the light of data derived from the cichlid flocks of other lakes. Particular consideration is given to the question of why the Cichlidae, especially members of the *Haplochromis* generic complex, have undergone such markedly explosive and adaptively multiradiate evolution whereas other groups have not reacted in the same way.

Finally, the evolutionary patterns seen in the *Haplochromis* species of Lake Victoria (and other African lakes) are reviewed in the context of evolutionary phenomena in general.

INTRODUCTION

THE freshwater fishes of the Great Lakes of Africa have long fascinated and perplexed students of evolution. As far back as 1913, Plate suggested that the cichlid fishes in Lake Tanganyika might show a mode of speciation different from that of the usual kind. This view, broadened to include fishes from other lakes, was supported by many subsequent workers, with Rensch (1933) a notable exception, and is still echoed today (Trewavas, Green & Corbet, 1972).

The problem has, if anything, increased in its complexity (and interest) since Plate's time, despite our greatly increased knowledge of these fishes. The publication recently of a book some 500 pages long (Fryer & Iles, 1972) summarizing researches into the biology and evolution of but one family of African fishes (the Cichlidae) is a good indicator of the problem's size.

Attention has always centred on species of the family Cichlidae, perch-like fishes widely distributed in tropical America and Africa, but also occurring in India, Shri Lanka (Ceylon) and Malagasi. This focus on African cichlids has, to a certain extent, distracted attention from other African freshwater fishes, and even from the cichlids of South America (see Lowe-McConnell, 1969), a sad but understandable state of affairs.

Each of the larger African lakes is characterized by a strongly endemic cichlid species complex (Table I). In most lakes these species dominate the fish fauna, both in an ecological and a taxonomic sense. Ecologically, the lake cichlids generally show a wider spectrum of adaptive radiation than do all the other families combined.

The bare figures of Table I show clearly the high degree of endemism, the extensive speciation undergone by these fishes and also the preponderance of cichlid taxa over those of all other families. Some measure of the morphological differentiation within the various species flocks may be taken from the number of endemic genera recognized in each lake.

Much of the literature dealing with African lake fishes has been concerned with these various phenomena, but in particular with the mode or modes of speciation involved, and the reasons for the disparate levels of taxonomic diversity found in different lakes (see Fryer & Iles, 1972). Far less attention has been given to the question of why the Cichlidae more than any other family display this evolutionary potential, a point to which I shall return later (p. 103).

One peculiarity of the African Cichlidae which Table I does not bring out is the contrast in diversity and numbers between lake and river dwelling species. The situation in most rivers is quite the reverse of that in the lakes, with species from other families providing the dominant faunal elements. A comparison between Lake Malawi and the Congo river illustrates this point very clearly, especially since both ecosystems have the most speciose fish faunas known from Africa.

In Lake Malawi there are *ca* 200 cichlid species comprising some 78 per cent of the entire fish fauna (*ca* 242 species), but in the Congo only 7 per cent of the 410 known species are cichlids (Lowe-McConnell, 1969, where comparisons between other African lake and river faunas are clearly demonstrated in text-fig. 2, and comparisons are also made between the African and South American faunas).

The contrast between cichlid and non-cichlid species is even greater if one compares Lake Victoria (with *ca* 170 species of Cichlidae) to any of its affluent rivers. In these, only three cichlid species have been recorded at any distance from the lake itself.

Obviously this peculiarity is closely connected with the question of what factors influence and stimulate speciation within the lakes. It is also, of course, associated with the greater number of ecological niches provided by a lacustrine environment. (Niche used here, and elsewhere, in the sense of Elton [1928] rather than that of Weatherley [1963].) These questions will not be considered further and are only mentioned because of their relevance to that of the ancestral species which must have populated the embryo lakes from the preexisting river systems.

To return to the lake cichlids. When the evolutionary radiation summarized in Table I is seen against a time scale for lake histories, its speed is outstanding. Dating

Table I

Cichlidae

Lake	Number of species: Total	Number of species: Endemic	Number of species: *Haplochromis* spp.	Number of species: Endemic *Haplochromis* spp.	Number of genera: Total	Number of genera: Endemic
Victoria*	*ca* 150–170	All but 3	All but 8	All but 1	8	4
Edward–George	*ca* 35–40	All but 5	All but 4	All but 1	4	0
Albert	10	4	6	4	2	0
Rudolf	7	2	2	2	3	0
Malawi	*ca* 200	All but 4	> 120	All but 1	23	20
Tanganyika	126	126	2	2	37	33
Nabugabo	10	5	6	5	4	0

Other families

Lake	Number of families	Number of species: Total	Number of species: Endemic	Number of genera: Total	Number of genera: Endemic
Victoria†	11	38	16	20	1
Edward–George	7	17	2	10	0
Albert	13	36	3	21	0
Rudolf	14	32	5	22	0
Malawi	8 + 2 ?	42	26	19	0
Tanganyika	13	67	47	29	7
Nabugabo	9	14	0	11	0

The total number of cichlid species and genera, together with the number of *Haplochromis* species, and the total number of non-cichlid families, genera and species, in six major (and one minor) African lakes. Only species actually recorded from a lake (i.e. excluding its affluent rivers) are registered. Data derived from various sources; estimates for Lakes Victoria, Rudolf, Edward and George based on personal observations (and Greenwood, 1973a for Lake George), those for Lake Malawi from Fryer & Iles (1972) and D. Eccles (personal communication).

* Excluding Lake Kioga and the Victoria Nile (although the inclusion of these water-bodies would not change the figures quoted).

† Excluding Lake Kioga.

of the lake basins is still far from settled, but current estimates suggest that the oldest (Lake Tanganyika) is not more than 10 million years old, that Lake Malawi is about 2 million years, and that Lake Victoria originated during the mid-Pleistocene, about 750 000 years B.P. (see summaries in Fryer & Iles, 1972 ; and Temple, 1969, for Lake Victoria). Lake age must undoubtedly be one of the factors affecting the degree of differentiation within a flock, as must the time and completeness of a lake's isolation from other major water bodies, be they lake or river systems. Thus the specifically depauperate and ecologically impoverished species flocks of Lakes Rudolf and Albert are a measure of age and lack of isolation. Both lakes are moderately young (early Pleistocene) and both have, or have had until recently, free contact with the Nile.

Lake Victoria, the lacustrine background to this essay, is younger, but it has been a closed drainage basin for most of its existence (Doornkamp & Temple, 1966 ; Temple, 1969 ; Greenwood, 1973a). Its cichlid flock is adaptively multiradiate, is highly speciose, but is composed of species which, with seven exceptions (two species of *Tilapia*, and five monotypic genera) are all members of a single genus, *Haplochromis* (but see p. 99). The five monotypic genera are all derived from and closely related to the genus *Haplochromis* (Greenwood, 1956a, 1959a ; also p. 99 below). In this respect its species flock can be considered a relatively simple one when compared with those of Lakes Tanganyika and Malawi (see Table I), as it can also be from the viewpoint of morphological divergence among its *Haplochromis* species.

The situation in Lake Victoria can probably be taken as a real model of a stage through which the flocks of Tanganyika and Malawi have already passed. Its importance in this context is enhanced by the fact that the components of the flock are already well-advanced along different paths of ecological specialization.

The comparison with Lake Tanganyika is perhaps less complete because there the species seem to be derived from two major phyletic components of the African Cichlidae (the so-called '*Tilapia*' and '*Haplochromis*' lineages ; see Regan, 1920 ; also Fryer & Iles, 1972). In Lakes Malawi and Victoria by contrast, the flocks, except for a small group of *Tilapia* species, are members of the '*Haplochromis*' lineage.

As was noted earlier, most attention has been paid to the problems of speciation within the lakes, and to purely taxonomic and ecological studies of the fishes. Little has been written about the phylogeny of a particular flock (Fryer's [1959] analysis of the 'Mbuna' generic complex in Lake Malawi excepted), and virtually nothing has been said about the evolution of adaptive characters within a flock. By the latter I mean particularly the anatomical basis for the adaptive trends observed within a phylogenetic framework.

I believe that the *Haplochromis* species flock of Lake Victoria provides very suitable material for such an analysis, the more especially since it can be linked with a fairly certain physiographical background to speciation within the developing lake basin. I also believe that from such an analysis, it is possible to throw light on one aspect of the question of why cichlids are able to undergo rapid and ecologically successful adaptive radiations when other families remain, by comparison, evolutionarily inert.

LAKE VICTORIA, ITS PHYSIOGRAPHY AND HISTORY

Lake Victoria is a large, and by comparison with other African lakes, shallow water body lying across the equator (0°21'N – 3°0'S, 31°39' – 34°53'E). Its surface area is approximately 69 000 km^2, its greatest length and breadth *ca* 400 and 320 km respectively. The coastline is extremely irregular, and totals some 3300 km in length. Especially in the northern half there are a number of large islands (Text-fig 1) whose shorelines are as varied as those of the mainland.

Broadly speaking, the habitat types provided by the shore comprise deeply indented, shallow and protected bays, sandy exposed beaches and, occasionally, rocky cliffs or broken rock exposures. Fringing papyrus swamps are common around much of the shore, and many of the bays terminate in broad swamps extending over several square kilometres.

Much of the lake is less than 20 m deep ; the deepest zone (60–90 m) lies somewhat eccentrically towards the eastern shore and occupies a kidney-shaped area (see Graham, 1929). The bottom profile in the deeper waters is not entirely flat, occasional 'hills' rise well above the general level of the lake floor. Much of the bottom in these deep areas is covered by a thick deposit of organic mud with, here and there, isolated patches of hard substrate (sand, shingle or rock). Organic mud substrata occur inshore as well, especially in sheltered bays, but also along the open coastline in protected regions. In most places where there is a sandy beach, the sand substrate grades imperceptibly into mud some few hundred metres from the shore ; rather rarely does the sand extend for more than two or three kilometres offshore and into water over 30 m deep. Substrate type appears to be an important factor in limiting the distribution of many species (see p. 46).

There is little annual variation in water temperature, the mean surface temperature being *ca* 24°C (that of deeper water about a degree lower). An annual cycle of thermal stratification (Fish, 1957 ; Talling, 1963) leads to a marked reduction of dissolved oxygen in deeper parts of the lake. The effects of this relative deoxygenation on the biology of fishes living in affected areas has not been satisfactorily investigated. At one time (Greenwood, 1965a) it was thought that no cichlids, and few other species, inhabited depths below about 30 m. However, later researches showed that many species of *Haplochromis* live in water at this depth and deeper. Indeed, there is probably no offshore area of the lake without its populations of *Haplochromis* species. What we have yet to discover are the reactions of these fishes during periods of deoxygenation, whether they can temporarily adapt their respiratory requirements to the new environment or whether there is some migration into shallower and better oxygenated zones.

In this respect Lake Victoria contrasts strongly with Lakes Malawi and Tanganyika. In these deep (704 and 1470 m) and trough-like lakes the lower water layers are permanently stratified (Beauchamp, 1964) and no fishes live there (Text-fig. 2). The greatest depth at which fishes have been recorded in Lake Tanganyika, for example, is about 200 m, and in Lake Malawi *ca* 100 m. But these deep-living species are exceptional, and most species (particularly Cichlidae) are restricted to the upper 30 or 40 m).

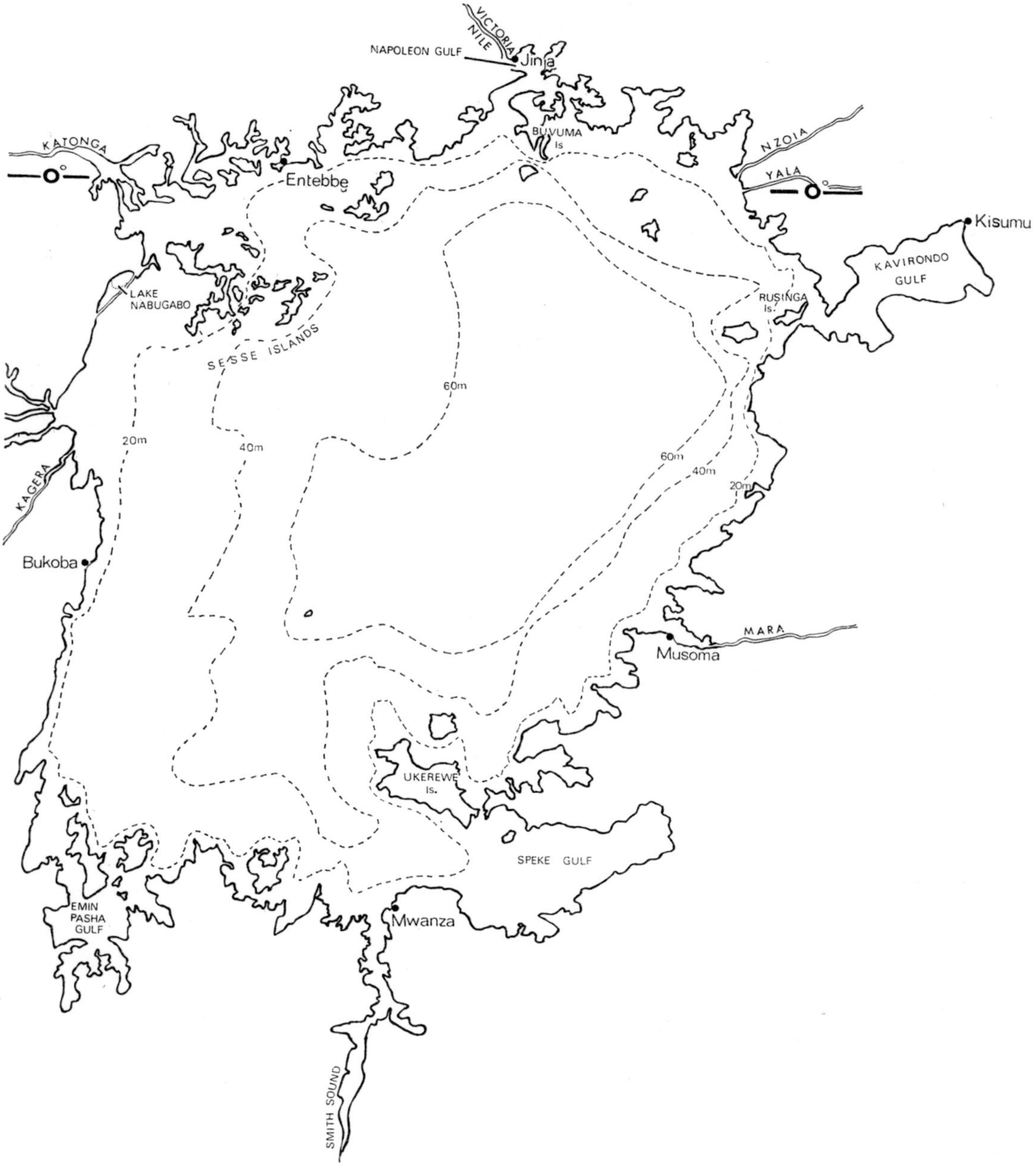

FIG. 1. Sketch map of Lake Victoria. The approximate positions of the 20, 40 and 60 m isobaths are indicated by dotted lines.

Detailed analyses of the chemical composition of the water are given by Talling & Talling (1965). Lake Victoria is categorized by these authors as a lake with a low total ionic concentration (conductivity between 90 and 145 μmho). There is virtually no short term change in major ionic composition.

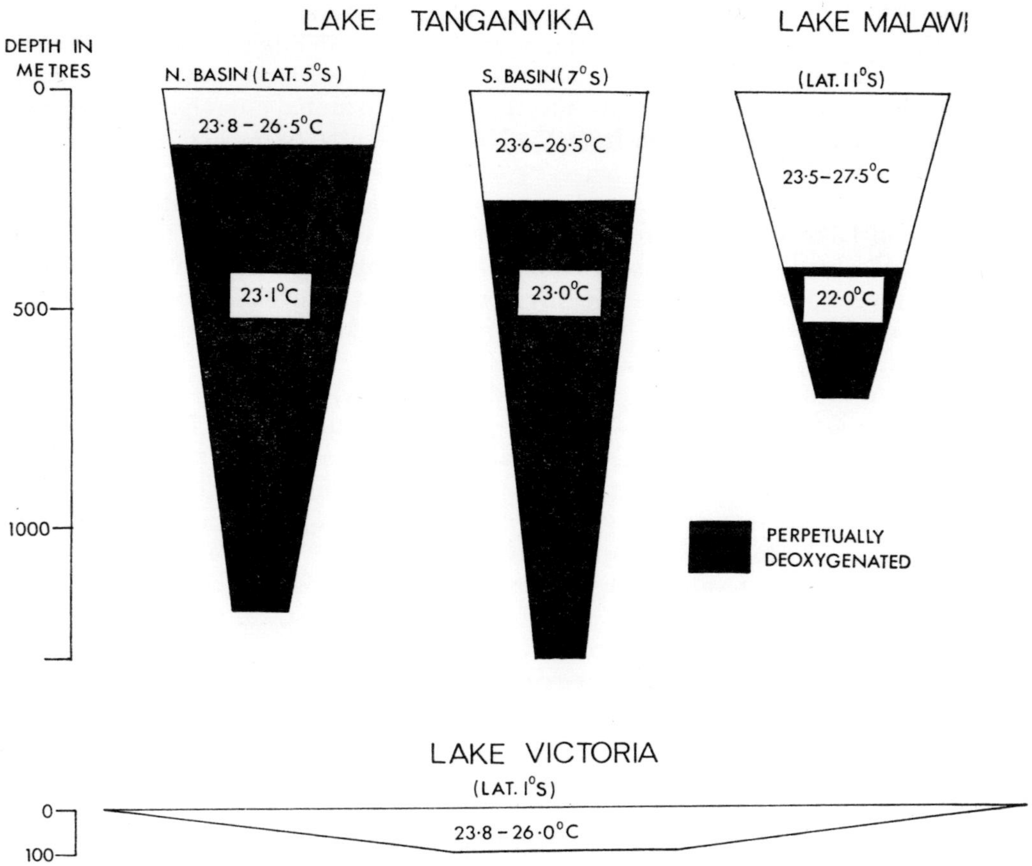

Fig. 2. Schematic cross-sections through Lakes Tanganyika (north and south basins), Malawi and Victoria to show differences in the shape and depth of their basins, and the extent of permanently deoxygenated water.

The horizontal scale is proportional for lake width at the latitudes indicated.

The annual temperature range of surface waters in the epilimnion is shown, as is the mean annual temperature of the hypolimnion (modified after Beauchamp, 1964).

Probably the major climatic factor affecting overall lake ecology is the biannual rainy season (most marked in the northern half of the lake), in November and December, and from March to May. Apart from any direct effect by rainfall on the water mass itself, the rainy seasons effect the inflow of water through not only the larger affluent rivers (Kagera and Nzoia) but also the numerous and often seasonal streams draining the surrounding countryside.

Lake Victoria is, and seemingly always has been, an internal drainage basin (Temple, 1969; Bishop, 1969; Kendall, 1969) fed principally by the four larger rivers (Kagera, Katonga, Nzoia and Mara). Its only outflow, formed late in the lake's history (Bishop, 1969), established contact with another, much smaller basin, that now occupied by the swampy Lake Kioga. Lake Kioga in turn has a river connection with the Nile at the northern end of Lake Albert. Faunistically, Lake Victoria is, however, isolated from the Victoria Nile and Lake Kioga by the now artificially submerged Ripon Falls. (The Owen Falls dam, built downstream of the Ripon Falls is probably an even more effective barrier to faunal interchange than were the falls.) Lake Kioga is cut off from the Nile by the utterly impassable Murchison Falls, some 40 m high.

Since the fishes of Lake Kioga, and especially the Cichlidae, are mostly conspecific with those of Lake Victoria (Greenwood, 1966a) it is very likely that the latter lake was the chief source of fish stocks for Kioga. Regrettably, the cichlids of Kioga have not been investigated at all thoroughly, but at least two endemic *Haplochromis* species occur there (Greenwood, 1967), and my own studies suggest that more await description.

Lake Victoria originated during the mid-Pleistocene, about 750 000 years B.P. (see Bishop, 1969; Doornkamp & Temple, 1966, for summaries of geological and other evidence). At that time the future lake basin was crossed by several westward flowing rivers, of which the Kagera, Katonga, Nzoia and Mara are present-day relicts. These rivers drained the eastern highlands of Kenya, and emptied into what is now the Congo system. A gradual but large-scale warping of the plateau surface between the two arms of the rift valley led to a reversal of river flow and a back-ponding of the western reaches of these rivers. It should be noted that, for some of the rivers, the western upwarp that interrupted and reversed their flows back towards the east was relatively slight, probably in the region of 30–50 m.

As a result of upwarping a two-way drainage was established, eastward into the developing Victoria basin, and westward into the proto-Lakes Albert and Edward. This drainage pattern still persists (Doornkamp & Temple, 1966; Greenwood, 1973a).

Temple (1969) is of the opinion that backponding of the reversed rivers began earliest in the southern region of the Victoria basin, and progressed northwards along the slowly sinking plateau. As a river valley gradually filled it became a shallow, dendritic lake. Lake Kioga, lying near the head of the now interrupted and reversed Kafu river valley, may well be a surviving example of such a lake (see especially text-fig. 4 in Bishop, 1969).

Eventually, each of the several lakes formed in this way overtopped the intervening and low watersheds and joined its neighbour. Developed and developing lakes were gradually linked together to form a single, expansive water body that occupied an area considerably greater than that of the present lake. Raised beaches also indicate that at one time the Pleistocene lake was considerably deeper than it is now.

Even after the single water body came into existence the lake basin was subject to periods of tectonic instability. These caused a tilting of the basin and correlated changes in water level at the raised and lowered areas of its margins. One important consequence of a major tilt to the northeast was the formation of an outlet (at Jinja) to the Kioga basin. Once the outlet was cut (possibly about 20–25 000 years B.P. ; see Bishop & Trendall, 1967 ; Bishop, 1969) there was a gradual fall in lake level to that of the present day.

It has been thought that during one period of the middle Pleistocene Lake Victoria was in contact, through rivers, with the proto-Lake Edward–George (Trewavas, 1933 ; Greenwood, 1959d, 1965a ; Fryer & Iles, 1972). Evidence for this connection stemmed mainly from overall similarities in the *Haplochromis* fauna of the two lake systems, and the supposed sharing between them of certain otherwise endemic species (Trewavas, 1933). Recent research on the *Haplochromis* of Lakes Edward and George (Greenwood, 1973a) suggests, however, that these resemblances are more likely the result of parallel evolution. The supposedly shared endemics are, in fact, specifically distinct and thus endemic each to its own basin.

Throughout the early history of Lake Victoria, and particularly during the later stages of tectonic stability, local and more widely spread climatic changes may well have produced fluctuations in lake levels if only by a few metres (Bishop, 1969 : 111 ; Kendall, 1969). Such changes would undoubtedly lead, first, to the formation of peripheral water bodies, and later to their reunion with the parent lake. A sequence of events like this would have a profound influence on the evolution, particularly the speciation pattern, of the fishes (see Greenwood, 1965a). I shall return to this point later.

But, even more important in interpreting the phyletic and evolutionary picture of the *Haplochromis* species flock, especially its adaptive radiation, were the stages of independent lakes through which the basin passed early in its history. In effect, the present lake must be considered an amalgam of several lakes (Greenwood, 1965a ; and p. 114 below).

THE FISHES OF LAKE VICTORIA

The non-cichlid fishes

Before going on to treat in detail the various elements of the *Haplochromis* species flock, some attention must be given to the other fishes with which the cichlids share their environment, and with which they may compete for food and living space.

Unfortunately it is not possible to discuss the intriguing and fundamental question of competition in any detail because too little precise information is available on the ecology of Lake Victoria fishes, both cichlid and non-cichlid. Likewise, little is known about the invertebrate animals on which so many of the fishes feed.

With a few partial exceptions, the 38 species of fishes belonging to families other than the Cichlidae share habitats with the cichlids (Greenwood, 1966a ; and p. 45 below). The partial exceptions are those species whose habitat ranges include zones of the lake where cichlids are absent. For example, certain airbreathing species like the lungfish *Protopterus aethiopicus*, the anabantoid *Ctenopoma murei* and at least two species of the catfish *Clarias* penetrate fairly deeply into papyrus swamps. Only one cichlid (*Hemihaplochromis multicolor*) is found in this habitat and even then it is confined to the peripheral and better oxygenated area. The small cyprinid *Engraulicypris argenteus* is the only truly pelagic fish in the lake. Inshore, where a distinction between pelagic and benthic zones is unrealistic, *Engraulicypris* does occur with several *Haplochromis* species, but it alone occupies the surface waters of the open lake.

Most non-cichlid species enter rivers and the larger permanent streams at all times of the year, but apparently few cichlids do so. Those that do (*Haplochromis nubilus, Hemihaplochromis multicolor* and *Astatoreochromis alluaudi*) are all species with a wide distribution in Uganda (Greenwood, 1959a, 1965b, 1973a).

The greatest intermingling of cichlid and non-cichlid species occurs in the littoral and sublittoral zones of the lake. Beyond a depth of *ca* 20 m cichlid (i.e. *Haplochromis* species) dominance is clear-cut, and few non-cichlids occur in deeper waters. Only one non-cichlid (the clariid catfish *Xenoclarias*) is confined to deeper water (10–90 m), a sharp contrast with the forty or more *Haplochromis* species known only from similar habitats. It is perhaps significant that, excepting *Xenoclarias*, non-cichlid fishes inhabiting deepwater areas are all from species whose adults reach a length of over 250 mm, usually over 500 mm.

Most non-cichlids do not breed in the lake, but migrate up streams and rivers to spawn during the rainy seasons (Whitehead, 1959 ; Greenwood, 1966a), again contrasting sharply with the cichlid species. There is, however, some evidence that a few non-cichlids are able to breed in particular lacustrine habitats (Corbet, 1960, 1961 ; the deepwater clariid *Xenoclarias* [Greenwood, 1958a] only recognized after Corbet's studies were completed, should also be included in this category). Interestingly, the majority of deepwater inhabiting non-cichlids (*Bagrus docmac, Mormyrus kannume, Synodontis victoriae* and *Xenoclarias eupogon*) are also those species thought capable of breeding within the lake. These species would, therefore, not have to undertake the extensive vertical and horizontal migrations necessary to reach suitable breeding sites in rivers. It would be interesting to know if the deepwater populations of these species still follow the seasonal breeding patterns of their shallow-water congeners. Of the deepwater non-cichlids not suspected of lacustrine spawning (*Protopterus aethiopicus* and *Clarias mossambicus*), one (*Clarias*) is already known to make long migrations from offshore regions to reach suitable spawning sites (personal observations). The breeding habits of *Protopterus* (nest construction, parental care etc., see Greenwood, 1958b) and aspects of its larval behaviour (Greenwood, *op. cit.*) certainly imply that this species is an obligatory inshore breeder.

Feeding habits of non-cichlid species have been studied in some detail (Corbet, 1961). The majority can be classified as insectivores, with chironomid larvae as their principal food organisms. Four species are essentially molluscivorous (feeding on both gastropods and bivalves) and three are predominantly piscivorous (with *Haplochromis* species as the main prey). The small pelagic cyprinid *Engraulicypris argenteus* mentioned before is the only species that can be considered a zooplankton feeder. Only two species, the characids *Alestes sadleri* and *A. jacksoni* are primarily herbivorous, feeding on rooted plants of the littoral region ; both species, however, also eat insects.

This résumé is perforce oversimplified; as Corbet (1961) cautions, the feeding habits of all the non-cichlid fishes are to a certain extent facultative (at least within and sometimes beyond the major food categories used here) and vary with the size of the fish. Despite the broadly overlapping food requirements of these fishes, Corbet's (*op. cit.*) elegant and detailed analysis led him to believe that there was little interspecific competition for food; to quote: 'The few species with specialized feeding habits appear to enjoy a superabundance of food, whereas the others achieve the same object by remaining mobile and facultative.'

Regrettably, there are no comparably detailed studies on the food of the Lake Victoria cichlids (see Greenwood, 1956–69). However, sufficient is known about these species to indicate the existence of a considerable and broad interspecific trophic overlap, as well as the existence among these fishes of trophic specializations not encountered in the non-cichlids (see pp. 30–44 below).

Corbet's work also shows that non-cichlid feeding habits in lacustrine and fluviatile environments are essentially similar. This fact could be of importance in the ultimate evolution of trophic diversity within the *Haplochromis* species flock (see p. 115). In rivers, the non-cichlid species are the dominant fishes (see p. 4 above). Thus the early colonizers of the embryo Lake Victoria (itself probably a series of river-like lakes) would be mainly non-cichlid species. In the present-day rivers of Uganda, for example, there are only one or two *Haplochromis* species (and two or three other cichlids) as compared to twenty or more non-cichlids. The restricted environment of a developing lake would probably favour fishes that could exploit unoccupied feeding niches. That the ancestral *Haplochromis* were able to respond to this selection is manifest in the trophic diversity of their descendant species. Some of the reasons for this cichlid potentiality are discussed in detail below (p. 103 *et seq.*).

The cichlid fishes

The exact number of cichlid species in Lake Victoria is still undetermined; every new collection yields undescribed species, especially now that fishery research vessels are operating in the deeper waters of the lake (Greenwood & Gee, 1969). It should also be noted, parenthetically, that cichlid species are being added to the fauna as the result of introductions from other lakes. In the following discussion these exotic species will not be considered; most belong to the genus *Tilapia* but inevitably species of other genera have been added unintentionally.

There are two endemic *Tilapia* species (*T. esculenta* and *T. variabilis*) and between 150 and 170 species of the genus *Haplochromis* of which all but one, *H. nubilus*, are endemic (see Greenwood, 1973a where the presumed occurrence of Victoria species in Lakes Edward and George is discussed fully, and the idea discounted).

Not all the *Haplochromis* species have been formally described, and the taxonomic revision of certain nominal taxa is still incomplete. The estimated number of *Haplochromis* species is based on collections still unworked, on personal observations in the field, and on information from fishery biologists now engaged on deep- and midwater trawling surveys.

In addition to the *Haplochromis* species *sensu stricto* there are four endemic monotypic genera (*Macropleurodus bicolor*, *Hoplotilapia retrodens* [Text-fig 72],

Platytaeniodus degeni [Text-fig 71] and *Paralabidochromis victoriae* ; see Greenwood, 1956a). All are derivatives of *Haplochromis* species. A fifth monotypic genus, *Astatoreochromis alluaudi*, again a *Haplochromis* derivative (Greenwood, 1959a, 1965c) is not restricted to Lake Victoria.

A sixth species, also of wide distribution outside the lake, is *Hemihaplochromis multicolor*. This species is probably the only cichlid in the lake that should more properly be considered a fluviatile than a lacustrine one. Until recently *Hemihaplochromis multicolor* was classified with *Haplochromis*. It is now separated from that genus because of its distinctive breeding habits (Wickler, 1963).

I have certain reservations about the phyletic soundness of recognizing, as genera distinct from related *Haplochromis* species, some of the endemic monotypic genera (see Greenwood, 1973a ; also p. 99 below).

At this point it is interesting to compare the cichlid species flocks of Lakes Victoria and Malawi. Lake Tanganyika is not brought into the comparison because its cichlid species flock (or more correctly, species flocks) is not dominated by *Haplochromis* and *Haplochromis* derivatives (see Regan, 1920 ; and Fryer & Iles, 1972 where Regan's views on the diphyletic origin of these fishes is questioned, I think justifiably on the evidence now available).

There are more *Haplochromis* species in Lake Victoria than in Lake Malawi (*ca* 150, *cf* 105) but far fewer endemic *Haplochromis*-group genera (four *cf* about 20 in Lake Malawi). Furthermore, these latter taxa in Lake Victoria are monotypic whereas those in Lake Malawi are mostly polyspecific (Trewavas, 1935 ; Fryer & Iles, 1972). What interpretation can be derived from these figures?

First, and most importantly, it should be stressed that the figures do truly represent differences between the degree of morphological differentiation existing in the species flocks of the two lakes. The phyletic conclusions to be drawn are less obvious (see p. 99). Possibly the greater age of Malawi and the history of its lake basin (a deep rift valley lake) are or were important factors, as are the different types of habitat provided by a 'graben' as opposed to a saucer-shaped ultimate basin. All in all it seems likely that the Lake Victoria *Haplochromis* flock (including all but one monotypic genus) can be looked upon, at this point in time, as an arrested early stage in the more complex and morphologically more differentiated type of flock seen in Lake Malawi. I believe that the multiple-lake origin of Victoria (see p. 114) combined with its youth (and perhaps the relative rapidity with which it passed through the multiple-lake stage) are reasons why there has been greater speciation and less opportunity for more profound morphological differentiation among the flock.

In certain respects this question of different degrees of morphological divergence between the two flocks is more apparent than real. Considering both flocks from a phyletic viewpoint (i.e. one where propinquity of descent is a more realistic yardstick for measuring taxonomic relationships than are morphological gaps *per se*), I can find few grounds for elevating some of the Lake Malawi taxa to generic rank. Nevertheless, there are still some Malawian taxa showing greater anatomical specialization than is seen in Lake Victoria. The Lake Victoria flock is undoubtedly a simpler one with a lower overall level of morphological divergence from the basic fluviatile

Haplochromis type (as represented today in East Africa by *Haplochromis bloyeti* [Greenwood, 1971]).

The two endemic *Tilapia* species of Lake Victoria (*T. esculenta* and *T. variabilis*) also provide a simpler picture, taxonomically and ecologically, than do the five endemic species of Lake Malawi (Lowe, 1952, 1953). There are good grounds for thinking that the Lake Victoria species were each derived from different ancestral lineages. Thus, strictly speaking, they do not constitute a species flock like the *Tilapia* of Lake Malawi (Fryer & Iles, 1972).

Both Victoria species are specialized phytoplankton feeders (Greenwood, 1953). *Tilapia esculenta* obtains its food principally from phytoplankton in suspension, but *T. variabilis* feeds mainly on the moribund phytoplankton of the bottom deposits. At least partly correlated with the trophic differences are marked interspecific differences in habitat preference (and in breeding biology). Both species are essentially from inshore regions, with *T. esculenta* penetrating into deeper water (Gee, 1968), but rarely to depths greater than 30 m. Although the species can occupy similar habitats, *T. esculenta* is commoner in sheltered gulfs and bays, and *T. variabilis* on exposed shores. (The principal references to the biology of *Tilapia* species in Lake Victoria are Lowe-McConnell, 1956; Fish, 1951, 1955; Garrod, 1957; and Fryer, 1961.)

As specialized phytoplankton feeders the two *Tilapia* species probably occupy a virtually unique trophic niche in the lake. The records of *Clarias mossambicus* having ingested large quantities of phytoplankton (Graham, 1929; Greenwood, 1966a) only reflect the omnivoracity of this species (Corbet, 1961) if, that is, the ingestion of phytoplankton is not just accidental.

Among the *Haplochromis* species, at least four combine the characteristic gut morphology of a vegetarian with known records of feeding on phytoplankton (see p. 39 below). None, however, shows the specialized pharyngeal dentition of the *Tilapia* species, nor such relative elongation of the gut. It seems likely that all four *Haplochromis* species get their food mainly from bottom deposits (i.e. are like *T. variabilis* in their feeding habits, but not in habitat as none has been captured in an exposed locality).

Only these four *Haplochromis* can be considered in any way trophically competitive with the endemic *Tilapia* species. The whole question of interspecific relationships between the *Tilapia* and other species of Lake Victoria is, however, insufficiently studied for there to be any clear-cut indication of their pattern or consequences. About all that can be said is that *Tilapia* and *Haplochromis* species occur together in most habitats, and at all sizes, and that some *Haplochromis* tap the same food sources as does *Tilapia variabilis*.

My recognition of *Tilapia esculenta* and *T. variabilis* (and by implication other members of the genus) as specialized species contradicts somewhat the views of Fryer & Iles (1969, 1972). These authors (1972) believe that species of *Tilapia* '. . . are in many respects more generalized than most fishes with which we are concerned (other lacustrine cichlids), and even their specialized feeding habits are little removed from those of "bottom grubbers"'.

That *Tilapia* are highly adaptable to a variety of habitats and environmental conditions (a phenomenon well established by field and experimental observations) might, I agree, be considered an indication of generalization. Stenotopic species usually evolve from a more generalized ancestor to exploit a particular niche. But, the pharyngeal apparatus of *Tilapia* (especially in the phytoplankton feeders), its musculature and dentition, and the alimentary tract modifications (see Greenwood, 1953) also constitute a specialized condition within the Cichlidae. Attention may be drawn to the differences between these characters in the phytoplankton-feeding *Haplochromis* species and those in the various *Tilapia* species. The former only show marked departure from the generalized, omnivorous *Haplochromis* in the somewhat lengthened gut ; the pharyngeal apparatus is nowhere near as specialized as it is in *Tilapia.* I would agree that, anatomically, *H. erythrocephalus* and other phytoplankton eating *Haplochromis* in Lake Victoria do not depart greatly from generalized 'bottom grubbers', but *Tilapia* I must consider specialized in its feeding habits.

Unfortunately, few comparative data are available for any intergeneric differences in digestive physiology. Recent work by Moriarty and Moriarty (1973) does not, however, indicate any differences in the ability to digest and assimilate blue-green algae between *Tilapia nilotica* and the phytoplankton-feeding *Haplochromis nigripinnis* of Lake George. (*Haplochromis nigripinnis* shows about the same departure from a generalized *Haplochromis* as does its trophic counterpart, *H. erythrocephalus*, in Lake Victoria.)

If it be accepted that the pharyngeal apparatus, and as a correlate the feeding habits, of *Tilapia* are specialized it is not surprising that this genus has failed to produce any great radiation in trophic adaptations comparable with that seen in the *Haplochromis* species flocks. The anatomically generalized fluviatile *Haplochromis* species (like other generalized animals) have a greater evolutionary potential, as will be discussed later (p. 103 *et seq*).

What is surprising, is the relatively low level of speciation in *Tilapia*. That there are about 70 species of *Tilapia* in Africa compared to nearly 300 *Haplochromis* species in Lakes Malawi and Victoria alone underlines this point. In Lake Victoria there are only two *Tilapia* species (each derived from a different ancestral lineage), and in Lake Malawi merely five species (but nevertheless the largest single natural assemblage of *Tilapia* species in any lake). Furthermore, among the fluviatile *Tilapia* species there are many with a considerably greater geographical range than any fluviatile species of *Haplochromis*.

The question of why *Tiplapia* did not speciate to the same degree as did the *Haplochromis* in Lake Victoria (whose developmental history [see pp. 10 & 114] would seem to provide an ideal background for allopatric speciation) is particularly difficult to answer. There are, among the *Haplochromis* of this lake, several instances of species multiplication within any one trophic group. That is, evolutionary change not involving any change in the particular trophic specializations possessed by the group.

The endemic *Haplochromis* species of Lake Nabugabo (a but recently isolated bay of Lake Victoria ; see Greenwood 1965b) are another example of speciation without noticeable adaptive change in feeding habits. The *Tilapia* isolated in this lake did not speciate.

It is possible that conditions in the various water bodies which have contributed to, and have been part of, Lake Victoria did not provide sufficient food for isolated *Tilapia* populations to survive long enough for speciation to take place. This, however, seems a most unlikely explanation. *Tilapia,* of several species, kept in aquaria and dams are able to survive and breed on the most atypical diets (including carnivorous ones), and their adaptability to adverse environments is well known.

About the only conclusion to be drawn at present is that some inexplicable genetical stability, and thus evolutionary conservatism, is an inherent characteristic of *Tilapia.* By contrast, *Haplochromis* appears to possess an extreme degree of genetical instability that manifests itself in the repeated explosive speciation of this taxon.

The theoretical resolution of this problem is not, unfortunately, quite so simple when it is extended beyond the genus *Tilapia* to include other parts of the tilapiine* lineage.

The supposedly tilapiine derivatives that feature so prominently in the species flocks of Lake Tanganyika (Regan 1920 ; Poll 1956) seem to show an evolutionary potential at least equal to that of the haplochromines in Lakes Victoria and Malawi and, indeed, in Lake Tanganyika. Doubt has recently been cast on the phyletic integrity of this lineage in Lake Tanganyika (Fryer & Iles, 1972). If the revised and admittedly tentative phylogeny of Tanganyika Cichlidae proposed by Fryer & Iles (*op. cit.* : 502–508) is a better approximation to reality than that implicit in Regan's (1920) classification, then the evolutionary potentiality of haplochromines is even more outstanding.

There is, however, one outstanding example of what seems to be a truly tilapiine radiation, the fishes of Lake Barombi-Mbo (Trewavas, 1972). The 11 endemic cichlid species of this isolated crater lake in northwest Cameroons have been carefully investigated by Trewavas (*op. cit.*). Four of the five genera present are endemic. Despite their peculiar oral and pharyngeal dentitions, and their unusual body form, the tilapiine affinities of the group are more obvious than are those of most Tanganyika genera. The endemic species of the single non-endemic genus (*Saroherodon*) depart quite considerably from the 'typical' species of that genus.

For the moment I would consider that the Barombi-Mbo cichlids provide the most serious challenge to the arguments against tilapiine affinities for the Tanganyika genera mentioned above.

The deviation, from their nearest relatives, shown by the Barombi-Mbo fishes is certainly greater than that seen in the four endemic monotypic genera of Lake Victoria. All four are undoubtedly haplochromine derivatives (Greenwood, 1956a) as is the fifth and more widely distributed taxon, *Astatoreochromis alluaudi* (Greenwood, 1956a, 1959a, 1965b 1973a).

That none of these Lake Victoria taxa is polyspecific like so many of their morpho-equivalents in Lake Malawi (e.g. the nine genera and 28 species of the 'Mbuna' complex, Trewavas, 1935 ; Fryer, 1959) may be attributable in part to the greater

* The terms 'tilapiine' and 'haplochromine' are used here without any formal taxonomic (i.e. subfamilial) connotation, and merely as convenient handles for the supposed two major cichlid lineages.

age of Malawi, and in part to the fact that the Victoria species have not exploited a habitat which was unexploitable by less specialized *Haplochromis* species.

The five monotypic genera have been distinguished from *Haplochromis* purely on the basis of the morphological gap separating each from any *Haplochromis* in the lake (Greenwood, 1956a). One genus, *Macropleurodus bicolor*, is, however, very closely similar to an extant *Haplochromis* species, *H. prodromus* (Greenwood, 1957 ; and p. 71 below) ; had more been known about the morphology of Victoria *Haplochromis* at the time of its description (and its later redescription), *Macropleurodus* might well have been included in that genus. The same reasoning applies to the classification of *Paralabidochromis victoriae*. This species (Greenwood, 1956a) is known only from a single specimen, whose dentition closely resembles that of the genus *Labidochromis* from Lake Malawi (Trewavas, 1935), a fact that certainly influenced my decision to place it in a separate genus.

The other two genera, *Platytaeniodus degeni* and *Hoplotilapia retrodens*, do have very distinctive dental characters (Text-figs 73 and 74 respectively ; also Greenwood, 1956a). Even now, neither can be related to any particular *Haplochromis* species within or without the lake. They provide, in fact, the only recorded instance of specialized dental or other characters not linked by intermediates to the generalized condition.

Astatoreochromis alluaudi is the only monotypic genus differentiated from *Haplochromis* by non-dental characters ; its differential characteristics are the higher number of dorsal and anal fin spines, especially the latter (four to six spines *cf* three in *Haplochromis*). It also differs from the other genera in having a distribution that extends beyond Lake Victoria to include Lakes Nabugabo, Edward and George, and several small lakes in western Uganda (Greenwood, 1959a, 1965b, 1973a). Unlike most Lake Victoria cichlids, *A. alluaudi* freely enters streams and swamps. Within the lake the species shows a wider habitat tolerance than do the other monotypic genera and, indeed, many endemic *Haplochromis* species of the inshore lake regions. But, like these various species it does not penetrate into water more than about 20 m deep.

The specialized crushing pharyngeal apparatus of *A. alluaudi* is identical with that found in two endemic *Haplochromis* species, *H. ishmaeli* and *H. pharyngomylus* (Text-fig. 5). The diet of all three species is identical, *viz.* gastropod molluscs, especially *Melanoides tuberculata*. Despite this shared pharyngeal specialization I do not think that *A. alluaudi* is closely related to either of the *Haplochromis* species. Indeed, there are good reasons for considering it to be derived from a different stem to that of the other monotypic genera and the rest of the *Haplochromis* species flock (Greenwood, 1954, 1959 ; also p. 100 below).

Trophically, all the monotypic genera (except *Paralabidochromis* whose feeding habits are unknown) can be classed as mollusc eaters. *Macropleurodus bicolor* feeds on gastropods, its feeding methods are like those of *Haplochromis* species that remove the snail from its shell before ingesting the soft parts (Greenwood, 1957, and p. 37 below). *Hoplotilapia* eats, principally, bivalves, crushing the shells between its broad bands of jaw teeth ; Text-fig. 74 (also Greenwood, *op. cit.*). *Astatoreochromis alluaudi* feeds on gastropods which are crushed in the hypertrophied pharyngeal mill (Greenwood, 1959a, 1965c). The feeding habits of *Platytaeniodus degeni* (Text-fig. 73) are largely unknown, but bivalves have been recorded from its gut contents, as have bottom detritus and insect remains (Greenwood, 1956a).

It is clear that despite the outstanding dental characters (Text-figs 73–75) of these genera, there is little to indicate that any one has entered an adaptive grade, or a particular habitat, which is not occupied by at least one *Haplochromis* species (see below). Perhaps the only point of interest in this context is to note that whereas *Haplochromis* species feeding on bivalves crush the shells by means of variously hypertrophied pharyngeal teeth, the monotypic genera achieve the same ends by using an hypertrophied oral dentition.

Four of the monotypic genera are female mouth brooders (as are all Lake Victoria *Haplochromis* species whose brood care is known) ; nothing is known about breeding in *Paralabidochromis victoriae*.

The monotypic genera of Lake Victoria can be equated with the 'Mbuna' generic complex of Lake Malawi (Trewavas, 1935 ; Fryer, 1959) on the basis of their having distinctive dental and oral characteristics. There, however, the comparison ends because the 'Mbuna' genera have occupied a habitat (rocky surfaces) to the virtual exclusion of *Haplochromis* species. In Lake Victoria the monotypic genera are, from the ecological viewpoint, indistinguishable from *Haplochromis* species.

THE *HAPLOCHROMIS* SPECIES FLOCK

Introduction

Having reviewed what might almost be considered the 'minor characters' amongst the fishes of Lake Victoria, attention can now be given to that ecologically and morphologically diverse, closely related and dominant assemblage, the *Haplochromis* species flock.

The estimated total of 150–170 *Haplochromis* species in Lake Victoria is based partly on the number of species already described (Greenwood, 1956–67 ; Greenwood & Gee, 1969) but also on the undescribed taxa from collections still under study and from information given to me by workers who have sampled areas of the lake in which I have not worked (particularly deep benthic and midwater habitats).

In the accounts that follow, data are drawn mainly from the 95 species that have been described or redescribed and which I have studied in the field. It must be emphasized that nothing learned from the still undescribed species in my possession seriously modifies this picture.

The term 'species flock' (= species swarm of Mayr, 1963) should, strictly speaking, be applied to a species assemblage of monophyletic origin. A monophyletic origin cannot definitely be established for the Lake Victoria species, although the evidence points in that direction. If the origin was not monophyletic then it was extremely oligophyletic (see below). The existing species are certainly more closely related to one another than to any species outside the lake, and it seems justifiable to refer to the assemblage as a species flock.

The *Haplochromis* of Lake Nabugabo (Greenwood, 1965b), of course, contradict these last remarks. But Nabugabo cannot, in this context, be considered 'another' lake ; were it not for a narrow sandbar it would be part of Lake Victoria.

A close relationship between the species of Lake Victoria and those of Lakes Edward and George (but not Rudolf and Albert) cannot be denied. In fact, if *Haplochromis* from either of these lakes were put into Lake Victoria, they would not seem at all 'out of place' to a taxonomist. The same cannot be said for the majority of *Haplochromis* species from Lake Malawi. For example, in all but one or two species the Malawi *Haplochromis* have the caudal fin covered by small scales; only the basal third, rarely the proximal half of this fin is scaled in Victoria species. And again, there are male breeding colours and colour patterns among Malawi *Haplochromis* and related genera that do not occur in the Victoria flock (Regan, 1921; Fryer & Iles, 1972; personal observations).

The problem of a mono- or polyphyletic origin for the Lake Victoria flock is not a simple one to solve (*Astatoreochromis alluaudi* excepted since it is manifestly more closely related to species outside the flock, see p. 100).

It has long been thought, on morphological evidence, that the flock could have stemmed from a single species (Regan, 1922; Trewavas, 1949). Finding evidence to refute this hypothesis is difficult, and is bound to be so if the mid-Pleistocene rivers of eastern Africa carried a similar *Haplochromis* species complement to that of the present-day rivers – one or at most two very closely related species.

There seem to be no grounds for assuming that the situation might be any different in mid-Pleistocene times. Even if different species did occur in neighbouring rivers, the chances are that they would be closely related. Thus it is not surprising to find that many elements from the species flocks of Lakes Victoria, Edward, George, Albert and Rudolf have a close overall resemblance to one another. The rivers that first drained into these lakes were all part of the east African highland drainage, and presumably carried the same or genetically similar *Haplochromis* species.

The relicts of this drainage system are now populated by a single *Haplochromis* species. Previously this fish was thought to be *H. wingatii* (see Trewavas, 1933) but recent work indicates that it is either *H. bloyeti* or a closely related species (Greenwood, 1971). *Haplochromis wingatii* is restricted to the Nile and Lake Albert, and appears to represent a lineage quite distinct from that of most east African lake *Haplochromis* species (see Greenwood, 1971, 1973a).

Lake Malawi is geographically far removed from the rivers of the old east African drainage and it is likely that its ancestral *Haplochromis* populations would differ from those of Lakes Victoria, Edward and Albert (see p. 99). The modern representative of the basal *Haplochromis* stock in Lake Malawi is thought to be *H. calipterus* (Trewavas, 1949). Structurally, this species too is not far removed from *H. bloyeti*.

Lake Tanganyika poses a problem. Geographically and hydrographically it lies within the range of present-day *H. bloyeti* (unpublished observations) yet its cichlid flocks are very different from those of Lake Victoria (Regan, 1920; Poll, 1956; Fryer & Iles, 1972). For one thing, the genus *Haplochromis* is barely represented in the lake by two species. One of these (*H. burtoni*) is a generalized species of restricted intralacustrine distribution (Poll, *op. cit.*; personal observations). The other (*H. horei*) is a moderately specialized predator also of relatively restricted intralacustrine distribution. Furthermore, nearly half the species flock is composed of genera apparently belonging to the tilapiine lineage (Regan, 1920; but see p. 17

above). Those genera belonging to the haplochromine lineage are as distinct from the general *Haplochromis* morphotype as are the monotypic genera of Lake Victoria, or even more so.

Here then we have the anomaly of two geographically distinct lakes, Victoria and Malawi, with basically similar species flocks neither of which resembles superficially or phylogenetically that of the geographically interposed Lake Tanganyika. No immediate explanation is available.

Of course the flock in Lake Tanganyika (even restricting comparisons to supposedly haplochromine derivatives) is much further differentiated than that of Victoria – probably a reflection of the lake's greater age – thereby making an inter-lake assessment of relationships very difficult. But, even allowing for differences in lake age, topography and history, there does seem to be a *prima facie* case for thinking that the ancestral species in Lake Tanganyika were quite different from the progenitor or progenitors of the Lake Victoria flock.

Basic morphology

Viewed in their entirety the Lake Victoria *Haplochromis* species show remarkably little diversity in body form, particularly when compared to the species flocks in other lakes (compare Text-fig. 3 with figs. 5–23 in Fryer & Iles, 1972).

The piscivorous predators, as a whole, are generally recognizable by their rather elongate form and large mouth (see Text-figs 3, 11–15, 52–56 and 58–64). But, in this trophic group there are several species (the '*serranus*' group in particular, Greenwood, 1967) that, apart from their larger size, closely resemble the trophically unspecialized species (*cf* Text-fig. 9 with Text-fig. 48). Even some of the most specialized predators, the paedophagous species (Greenwood, 1959b; and pp. 31–37 below), are not particularly outstanding in their superficial appearance (see Text-figs 16–18). There are, of course, exceptions and again most of these are found amongst the piscivores. In these species the outstanding morphological features are associated with the head, like a strongly prognathous lower jaw, an upwardly directed mouth, or the deeply concave dorsal head profile often associated with this mouth form (see Text-figs 14, 56, and 64).

The generalized *Haplochromis* body and head shape (found in more than half of the known species) cloaks a wide variety of trophic specializations, from algal grazing to mollusc eating and even predation on other *Haplochromis*.

These remarks apply to first impressions coming from a superficial examination of the fishes. Closer inspection in many cases shows differences in jaw and tooth morphology among otherwise similar species. Indeed, it is in the head and dentition that the real diversity of these species becomes apparent, and in which lies the evolutionary success of the *Haplochromis* species flock. Nevertheless, even the most extreme forms (except for two of the monotypic genera; p. 103) are linked with the generalized type by species showing between them all intermediate stages in the development of a particular characteristic.

From the last few remarks it must not be concluded that every species is immediately distinguishable by its cranial characters. Rather, these should be taken as 'species group' characters, shared in many instances by more than half a dozen

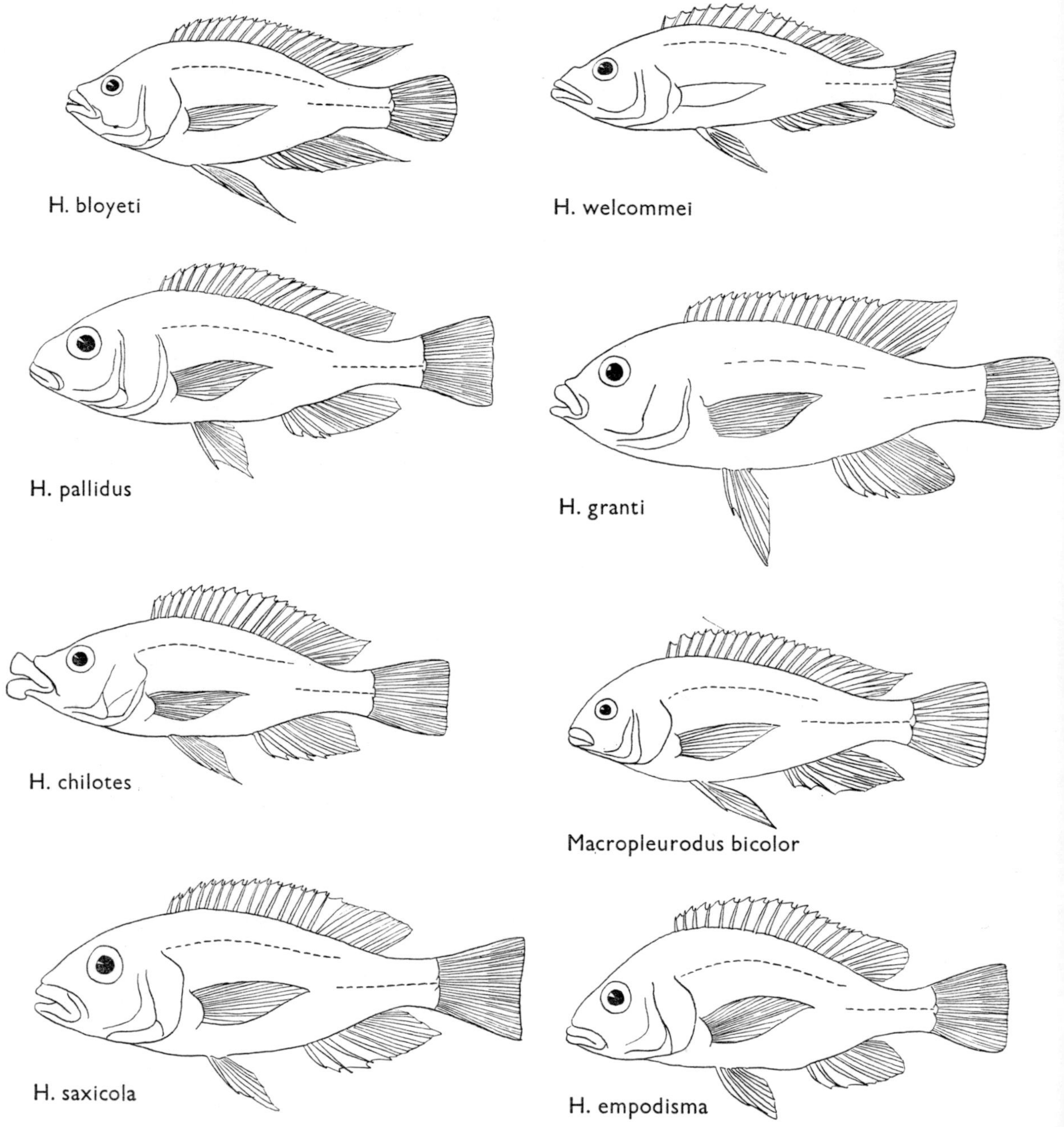

FIG. 3. Outline drawings (not to scale) of various Lake Victoria *Haplochromis* species (and one monotypic genus, *Macropleurodus bicolor*) to show range of body form in this species flock.

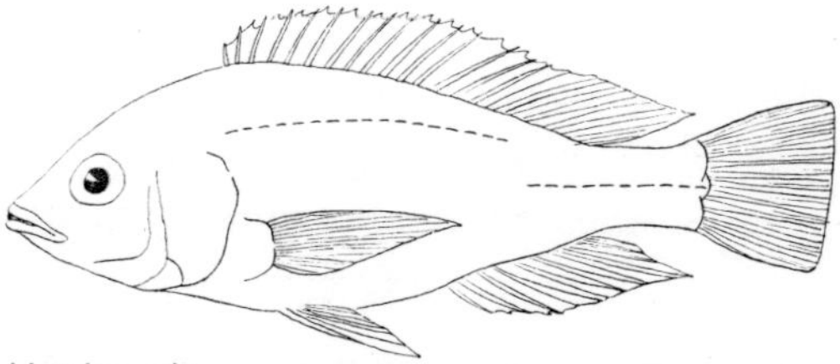

H. ishmaeli

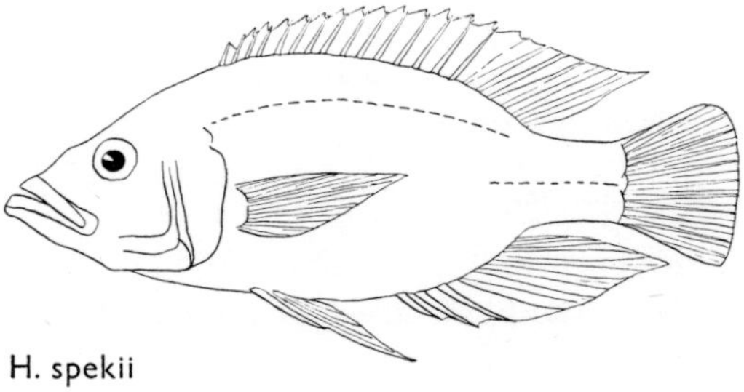

H. spekii

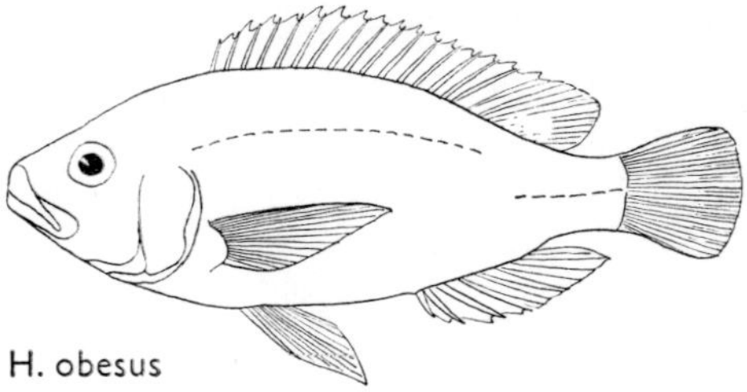

H. obesus

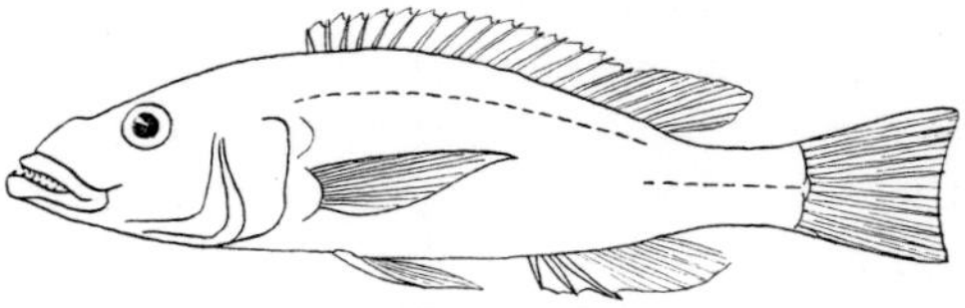

H. dentex

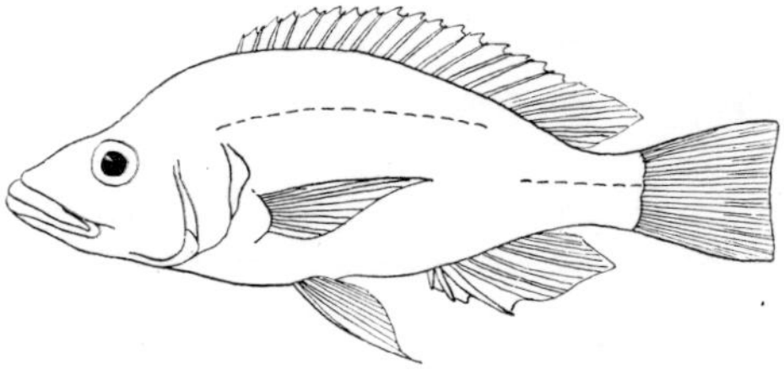

H. parvidens

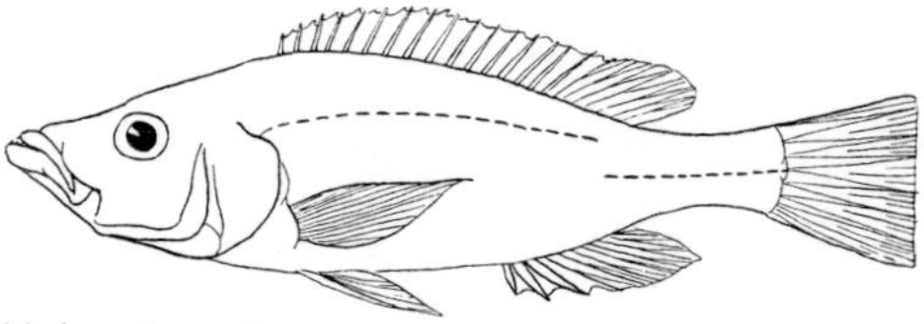

H. longirostris

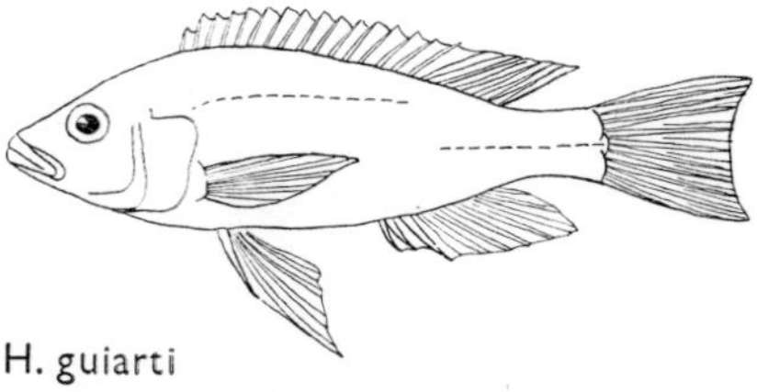

H. guiarti

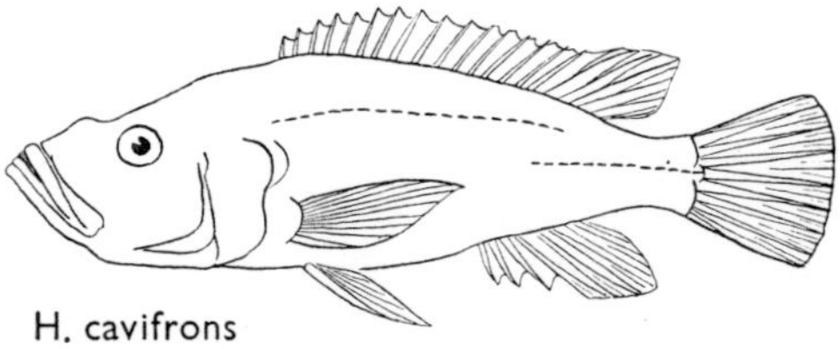

H. cavifrons

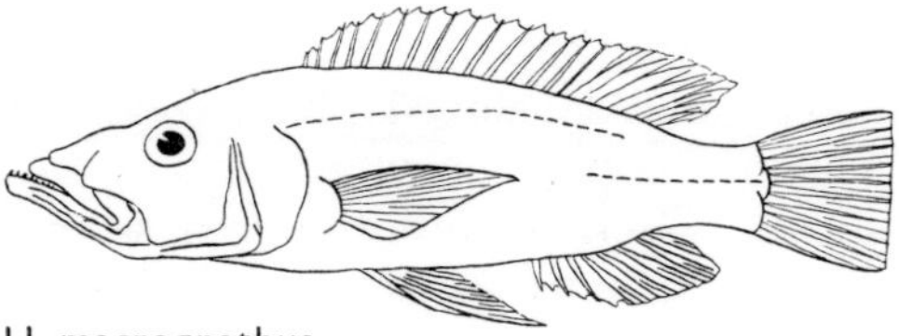

H. macrognathus

species. The two mollusc-eating species groups (Greenwood, 1957, 1960), the '*tridens*' group of species feeding on benthic Crustacea in deep water (Greenwood & Gee, 1969) and the several morphogroups of piscivorous predators (Greenwood, 1962, 1967) typifying this situation. Within a group the morphological characters separating species are mainly slight proportional differences, squamation patterns, differences in the number or disposition of teeth (and less commonly tooth shape) and, most clearly, differences in male breeding coloration (see p. 52 below and Plate I). In most of what may be termed 'specific characters', except male coloration, there is a high level of individual variability.

To exemplify this last point, and at the same time to emphasize the relative invariability of some morphological features, one may consider the meristic characters of the flock. The range of fin ray numbers, lateral line and most other scale counts, and the number of vertebrae is such that the range for the entire flock can be encountered in one species. There are a few exceptions to this generalization ; all are species with an elongate body form. In the exceptional species, the modal number and upper limits for the range of vertebral and lateral line scale counts may lie beyond the general range, but the lower limits lie within that range.

Few species show any external evidence of modifications associated with a particular habitat. Exceptional in this regard are some, but not all, species from deeper water (i.e. > 50 m). These fishes have relatively larger eyes, and the cephalic laterosensory canals and openings are slightly enlarged, presumably as adaptations to a dimly lit environment (Greenwood & Gee, 1969 ; Greenwood, 1973a).

Adult individuals of most species are small fishes, between 70 and 110 mm long, but with some of the piscivorous predators growing to lengths of 180–220 mm.

In all the morphological features discussed so far, the Lake Victoria *Haplochromis* flock (including the monotypic genera) shows less diversity than does the flock in Lake Malawi (even if the derivative genera are excluded). Because few morphometric data are available for individual species of Malawi *Haplochromis*, it is impossible to compare the levels of intraspecific variability in the two flocks.

Contrasting with the relative uniformity of body shape, the jaw teeth and dental patterns in Lake Victoria *Haplochromis* show a much wider range of diversity (see Text-figs 4, 36, 39 and 41). It is this particular diversity, more than any other factor, that has contributed to the success of these fishes in the lake (and in other lakes). There are, of course, correlated changes in syncranial architecture, especially neurocranial shape, and many of these are probably to be considered the primary changes involved in an adaptive radiation. Certainly, taken in concert, skull form and dentition are basic to the trophic radiation so characteristic of all cichlid species flocks.

The well-developed pharyngeal apparatus of the cichlids (i.e. the toothed and separate upper pharyngeal bones and the toothed but suturally united lower elements, together with the associated musculature) are effectively a second pair of jaws. In many species these 'jaws' are of greater importance than the true jaws, or at least have given the species an enhanced potential for exploiting a wider variety of food sources.

Examples of this potential realized are found in the phytoplankton feeders (Greenwood, 1953), the mollusc eaters (Greenwood, 1959a, 1960) and in the piscivorous

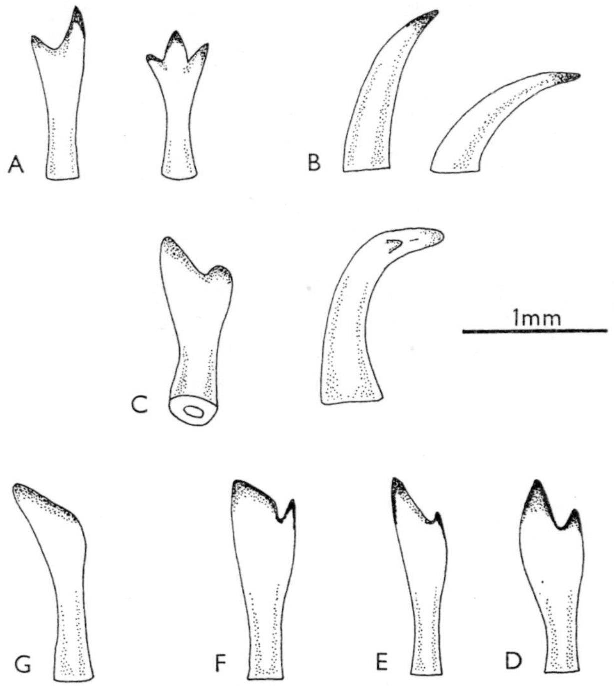

FIG. 4. Jaw teeth of various *Haplochromis* species.

A : *H. macrops* (an omnivore) ; bicuspid outer and tricuspid inner tooth from the dentary. These teeth may be considered as representative of the typical generalized tooth form.

B : Unicuspid, caniniform outer teeth from the premaxilla of *H. dentex* (a piscivore), showing different degrees of curvature.

C : Stout, strongly recurved, bicuspid outer row tooth from the dentary of *H. welcommei* (a scale scraper) in buccal (left) and lateral (right) views.

D–G : Teeth of four vegetarian species showing morphological stages in the evolution of specialized, obliquely cuspidate outer teeth in the periphyton grazer, *H. obliquidens* (G). The species represented are : (D) *H. erythrocephalus* (phytoplankton eater), (E) *H. nuchisquamulatus*, (F) *H. lividus* and (G) *H. obliquidens*, all periphyton grazers (see p. 39).

predators (Greenwood, 1962). Phytoplankton feeders have the pharyngeal teeth fine and hooked, thus enabling the fish to comb aggregates of mucus and phytoplankton into the oesophagus. In mollusc eaters the molariform pharyngeal teeth (Text-fig. 5), strong pharyngeal bones and powerful upper branchial musculature enable the fish to crush gastropod shells. In piscivores (Greenwood, 1962 : 211) there is a macerating action of the strong but fine upper and lower pharyngeal teeth moving against the prey caught between them. This allows the predator to ingest much larger prey than would be possible if the food was bolted whole (as is the usual way in most non-cichlids).

Such examples are, in some respects, extreme cases. Most species have what can be described as a 'general purpose' pharyngeal apparatus. Consequently there is, on the whole, rather less diversity in bone form and tooth shape than is seen in the morphology and dentition of the jaws.

In the jaws (premaxillary and dentary bones) there is always an outer row of large teeth, followed by from one to five (sometimes more, but usually two or three) rows of much smaller teeth (Text-fig. 41A). Teeth in the outer row show greater variety in shape and size (Text-fig. 4) than do those of the inner series.

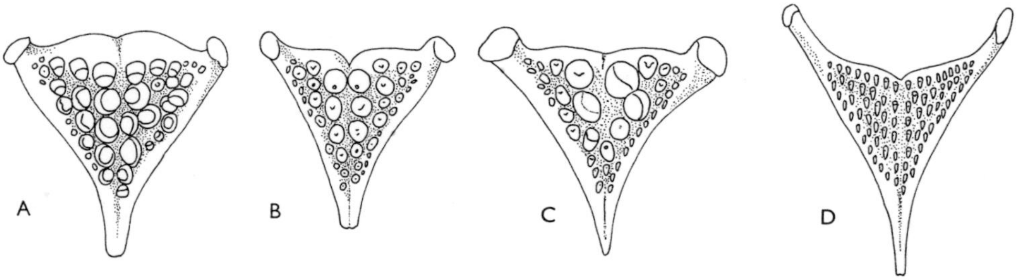

FIG. 5. Lower pharyngeal bones (in occlusal view) of four species to show increasing molarization of the teeth associated with an increasingly molluscivorous diet. *Haplochromis empodisma* (extreme right) has a mixed diet of insects and bottom detritus while *H. pharyngomylus* (extreme left) is exclusively molluscivorous when adult. Drawings not to scale. B: *H. obtusidens*. C: *H. humilior*.

Inner teeth are generally tricuspid or unicuspid ; their main contribution to dental diversity is in the number and pattern of the rows present. For example, there are the broad bands of fine, tricuspid inner teeth in those species that scrape epiphytic and epilithic algae (e.g. *H. obliquidens*, *H. lividus*, *H. nigricans* and *H. nuchisquamulatus* ; Greenwood, 1956b) or, like *H. welcommei*, which scrapes scales from the caudal fins of other cichlids (Text-fig. 39 ; and Greenwood, 1966b). Broad bands, this time of stouter teeth (Text-fig. 41), are found in certain of the species feeding on molluscs, either by wrenching the snail from its shell or by crushing the shell between the jaws as in *H. prodromus*, *H. granti* and *H. xenognathus* ; see Greenwood, 1957. The ultimate development of this trend is seen in the broad and posteriorly expanded inner tooth rows of *Hoplotilapia retrodens* and *Platytaeniodus degeni* (Text-figs 73 and 74).

Teeth from the outer row in the premaxilla and dentary are, despite their diversity in form, variants of two basic types (Text-fig. 4). In one, the crown is compressed and unequally bicuspid, in the other the crown is conical and protracted into a single sharp point (i.e. it is unicuspid). Occasionally, tricuspid teeth are found in the outer rows, but with few exceptions (e.g. *H. tridens* and related species ; Greenwood & Gee, 1969) are never common anteriorly in the jaws.

Teeth morphologically intermediate between the fully bicuspid and the unicuspid type are known, but never seem to occur in the same species or in the same individual (except as the result of wear on a bicuspid).

The bicuspid tooth is, by analogy with teeth in generalized and fluviatile species like *H. bloyeti*, taken to be the basic and generalized tooth form (Text-fig 30B). It also seems to be the juvenile tooth form irrespective of the definitive tooth shape for a particular species. This temporal succession is seen, albeit in muted form, among species whose definitive dentition is that of unequally bicuspid teeth ; in some large antividuals of these species there is often an admixture of bi- and unicuspids in the inderior and lateral parts of the dental arcade.

One or two relatively enlarged unicuspids are generally developed at the posterior end of the outer tooth row in the premaxilla, irrespective of tooth shape elsewhere in this jaw ; exceptional are the few species like *H. obliquidens* (Text-fig 4G) where crown form is very greatly modified from the usual condition.

Tooth succession is vertical, the replacement teeth developing in alveoli beneath the erupted and functional teeth. As far as can be determined, the replacement of outer teeth is more regular than that of inner ones (personal observations).

The basic bicuspid tooth has one cusp (the major cusp) noticeably larger than the other (Text-fig. 4A and D). Both major and minor cusps are triangular in outline, and lie in the same plane. Variations on this pattern involve changes in the size of the apical angle of the cusps, changes in the relative sizes of the two cusps or the inclination of one cusp away from the plane of the other. The minor can be as large as the major (Text-fig. 36) or the latter can be drawn out obliquely into a scraping blade many times larger than the minor cusp (Text-fig. 4G).

Another but rarer type of variation involves a coarsening of the whole tooth with a consequent increase in the breadth of the crown ; sometimes this trend is combined with one of those noted earlier, and results in the powerful teeth of *Macropleurodus bicolor*, an admittedly extreme example of this trend (Text-fig. 75).

The length of a tooth's neck also varies interspecifically, as does its circumference. In consequence, teeth can be short and stout, long and slender, or of a intermediate type, but all with a similar cusp shape.

Unicuspid teeth differ in relative size, degree of curvature (usually in the neck of the tooth, but sometimes in the angle between the crown and neck, as in *H. obesus*, *H. maxillaris* and *H. melanopterus* ; see Greenwood, 1959b), or in the angle at which the whole tooth is implanted on the jaw bone. These differences are well exemplified by the more specialized piscivorous species, and by some of the specialized insectivores and certain mollusc eaters. In the former group (Greenwood, 1962, 1967), the teeth are large and strongly recurved, thereby providing a means for gripping the prey while the pharyngeal teeth macerate it preparatory to swallowing. In certain specialized insectivores and mollusc eaters the teeth are also relatively enlarged, and although the tips are recurved, the teeth are implanted so as to project forward (Greenwood, 1957, and 1959b : 207–211). The procumbent teeth in the insectivores (especially *H. chilotes*, Text-fig. 33) create an effective 'forceps' used to remove burrowing insects from their holes. In the molluscivores, the teeth are used either to hold a snail while it is eased from its shell, or actually to crush the shell away from the body.

Lake Malawi *Haplochromis* (and especially the derivative genera) show an even greater range of dental morphology and arrangement. The difference is clearly seen by comparing Text-figs 4, 36 and 73–75 here with those for Malawi species published by Fryer (1959) and Fryer & Iles (1972) ; similarly for the haplochromine species of Lake Tanganyika (Poll, 1956 ; Fryer & Iles, *op. cit.*).

The whole subject of cranial and dental morphology in the Lake Victoria *Haplochromis* flock is dealt with in greater detail below (pp. 56–99). For the moment it must suffice to note that although species showing specialized dental, pharyngeal and syncranial characters are very distinctive (and readily advertise their feeding habits) the majority are linked, through species with these characters at an intermediate stage, to the generalized type (see Text-figs 4, 5 and 65–69).

Many of these morpho-lineages appear also to be truly phyletic ones because the component taxa can be related, primarily, though shared specializations, and secondarily by degrees of specialization in the same characters. There are a few instances where intraspecific variability in a complex of specialized characters is such that, were only the extreme individuals known, they would probably be classified as a separate genus (for example, *H. xenognathus*, Greenwood, 1957 ; and *H. welcommei*, Greenwood, 1966b).

Surprisingly, amongst species with such diverse feeding habits, there is little diversity in the shape, length or number of gill rakers; the modal numbers of rakers are 9 or 10. The plankton-feeding species (e.g. *H. erythrocephalus* ; Text-fig. 6) have

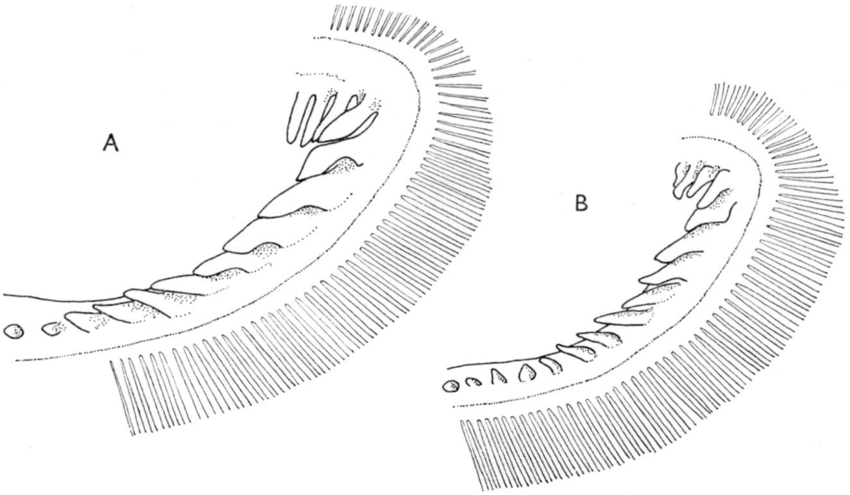

FIG. 6. Outer row of gill rakers on the first gill arch (left) of A : *H. brownae* (an insectivorous–piscivorous species) and B : *H. erythrocephalus* (a phytoplankton feeder).

rakers that are but marginally longer than those of an insectivore, are only a little more closely spaced, and are more numerous by one or two rakers. No explanation can be offered, except that the feeding mechanism of phytoplankton feeders may not require the evolution of close-set and fine rakers. These species entangle the plankton in mucus boli that are too large to pass through even the relatively wide spaces between rakers of the first row ; subsequent rows of rakers interdigitate and have even smaller interspaces (Greenwood, 1953 ; and unpublished observations on several *Haplochromis* species).

The theory of character displacement (Brown & Wilson, 1956) may be discussed here because it has been largely established on morphological characters, and because it would seem to be of particular relevance to a species flock situation. Its ecological counterpart, the concept of competitive exclusion (Gause's Law), will be considered later (p. 48) when discussing species' interrelationships in the lake.

In essence, the theory of character displacement postulates that closely related sympatric species will generally show more differences than similarly related allopatric ones. In other words, selection will sharpen differences between sympatric species if these differences lead to reduced interspecific competition. A species flock should, therefore, provide good material against which to check the first premise of the theory, although the second premise cannot be tested because none of the species ever occurs allopatrically with a close relative.

The general effect of character displacement seems to be discernible within the flock for many morphological features (and, by implication, even more clearly for ethological characters, see p. 51). A most striking example is provided by the differences in male breeding coloration (see p. 52). This example is particularly interesting because interspecific colour differences are most marked among species that occur syntopically and are less obvious between allotopic species.

The picture for other morphological features, especially anatomical ones, is far less obvious, and at first glance might even seem to contradict the idea of enhanced differentiation between sympatric species.

To illustrate this difficulty one can take the several anatomically and morphometrically similar species clustered around any one adaptive peak in a phyletic lineage. This is especially well demonstrated among such disparate groups as the generalized insectivores (Greenwood, 1960), in the specialized mollusc crushers (Greenwood, *op. cit.*), and particularly in the piscivores (Greenwood, 1962, 1967). However, traces of character displacement are evident, not so much at the level of individual species but at the level of species groups (i.e. phyletic lineages.) The characteristic features of these groups are, almost without exception, trenchantly defined.

Proximately, of course, what are now recognizable as phyletic groups must have originated as species. It was presumably at that level and at that time that selection pressure, and hence character displacement, was most intense.

The mosaic of interspecific similarities and dissimilarities in the *Haplochromis* of Lake Victoria could well mirror the way in which the flock evolved. Character displacement (and the origin of phyletic lines) would be most marked during early phases of lake development when a high premium might be placed on trophic specialization. Character replication, through simple speciation without obvious adaptive change, is likely to be a feature of later lake development, with the isolation and reincorporation of peripheral, Nabugabo-like lakes (see p. 112).

That intragroup character replication should exist on the scale it does among these fishes is certainly unusual (see the numerous contrary examples cited by Brown & Wilson, 1956). It seems to suggest two possible explanations, either a lowered level of competition and selection during certain phases of lake evolution, or the involvement of adaptive characters other than those reflected in morphological features. A temporal factor may also be involved. On a purely subjective assessment there seems to be greater interspecific differentiation among the *Haplochromis* species of Lake Malawi than those of Lake Victoria. The Malawi flock is older (see p. 6), and during its longer life perhaps some of the less distinctive taxa (i.e. less particularly specialized species) have been eliminated at times of more rigorous selection.

Feeding habits of the *Haplochromis* species

The importance of trophic specializations in the adaptive radiation of Lake Victoria *Haplochromis* species has long been recognized (see Regan, 1921, in the prefactory remarks to the first systematic revision of these fishes). The full magnitude of this radiation only became apparent, however, when field studies were made (Graham, 1929 ; Greenwood, 1959c, 1965a, 1973b, and in taxonomic papers dealing with these species, 1956–69).

Lake Victoria *Haplochromis* species are by no means unique in this respect, and the phenomenon can be considered a characteristic of cichlid fishes in many African lakes (and elsewhere in the world). Outlines of these different radiations were published, for Lake George by Greenwood, 1973a ; for Lake Malawi by Fryer, 1959 and especially by Fryer & Iles, 1972 ; for Lake Tanganyika by Poll, 1956 (also in Fryer & Iles, *op. cit.*), and for Lake Albert by Trewavas, 1938.

When discussing *Haplochromis* feeding habits in Lake Victoria certain points must be borne in mind. First, the food of small fishes (i.e. < 20 mm long) is unknown for the majority of species, partly because small individuals were rarely caught, but mainly because it is impossible to identify fishes of this size. Second, fishes caught by trawling in deep water usually have everted guts when brought to the surface, and much of the food is lost in this way. Finally, there has been little detailed research (like that of Corbet's [1961] on non-cichlids) into the food of *Haplochromis* ; what data there are, derive chiefly as an offshoot from my own taxonomic studies.

Within these limitations it is possible, however, to say that every major food source in the lake, except for zooplankton, has been exploited by one or several *Haplochromis* species (Text-fig. 7).

Insects, especially larval and pupal chironomids and Ephemeroptera, are probably the most important food organisms. Besides the 11 species* feeding principally on insects many others include some insect material in the diet, either regularly or opportunistically.

Most insectivores are morphologically (Text-figs 3, 8–9 and 42) and dentally generalized fishes, with bicuspid jaw teeth, generalized jaw structure (Text-figs 31

* Only the 95 species so far dealt with in my revision are used in these analyses.

FIG. 7. Diagrammatic representation of feeding habits in the Lake Victoria *Haplochromis* species studied to date (see p. 19). For details of species in the *H. macrops, H. serranus, H. prognathus* and *H. tridens* species complexes see text, pp. 58; 81; 85; and 67.

Where more than one food is represented, the lengths of the bars are not strictly proportional ; they should be considered merely as indicative of the relative importance of the particular foods in the diet of that species. Only the principal types of food organisms are shown. The species are arranged in phylogenetic groups (see Text-fig. 70).

and 33) and a pharyngeal apparatus as unspecialized as the primary jaws (Text-figs 5D and 30). The indications are that these species feed by sucking or picking their prey from the bottom or other substrata (plants, rocks, etc.).

The most specialized insectivore, *H. chilotes* (Text-fig. 10), is also, by virtue of its hypertrophied lips, the most easily recognized species in the whole flock. Anatomically, *H. chilotes* has a forceps dentition of procumbent, unicuspid and slightly curved teeth (Text-fig. 33 also see p. 27 and p. 61 below). With these teeth *H. chilotes* is able to extract the larvae of a boring mayfly (*Povilla adusta*) from their burrows ; prawns are also probably removed in the same way from cracks and holes in rocks. The hypertrophied lips (which are highly variable in their degree of development) seem to serve merely as shock absorbers when the fish forces its mouth against the rock or wood face ; histological examination of lip tissue does not show any particular increase in the number of sensory cells as compared to lip tissue from other species. Free-living insect larvae are also eaten by *H. chilotes*, and presumably taken, like the food of less specialized species, from the lake bottom.

Another species, *H. chromogynos*, has teeth and jaws closely similar to those of *H. chilotes* (see Text-fig. 33), but there is no suggestion from its known food (larval Diptera) of specialized feeding habits.

Next to insects, fishes are probably the most important source of food for *Haplochromis* (Greenwood, 1962, 1965a, 1967). I have estimated (Greenwood, 1959c, 1967) that about 40 per cent of *Haplochromis* species in Lake Victoria are piscivorous predators. With more knowledge of deep water species (Greenwood & Gee, 1969 ; unpublished observations) I would now reduce that figure to one of about 30 per cent, still a high proportion.

Most of the fishes eaten by the piscivores are other *Haplochromis*, although small cyprinids (especially *Engraulicypris argenteus*) constitute part of the diet in the larger predators. The fish remains recovered from a predator's guts are so macerated that it is generally impossible to identify the species of the *Haplochromis* prey consumed. Generic identification, however, is certain. No *Tilapia* remains have ever been identified with certainty ; probably this is because the young *Tilapia* (of both species) are 'protected' in their nursery zones (where *Haplochromis* species do not penetrate) until they reach a size where they are too large to be eaten by a predatory *Haplochromis*.

The clearly recognizable 'predatory' facies of fish-eating *Haplochromis* (see Text-figs 3, 12–15 and 52–63) was noted before. There are, however, a few species which superficially and even in their dentition and mouth size could well be mistaken for generalized insectivores (see Text-figs 48 and 50). The principal components of the 'predatory facies' are a large, moderately distensible mouth, strong unicuspid jaw teeth (Text-fig. 4B), and pharyngeal teeth, also strong, with the crowns so orientated and shaped as to form a macerating mill. Some piscivorous *Haplochromis* are among the largest in the lake, reaching, when adult, lengths of between 200 and 300 mm.

Also to be included in the piscivore category is a group of eight species that feed exclusively (or principally) on the embryos and yolk-sac larvae of other cichlids (Greenwood, 1959b, 1967). The feeding habits are definitely established for six of

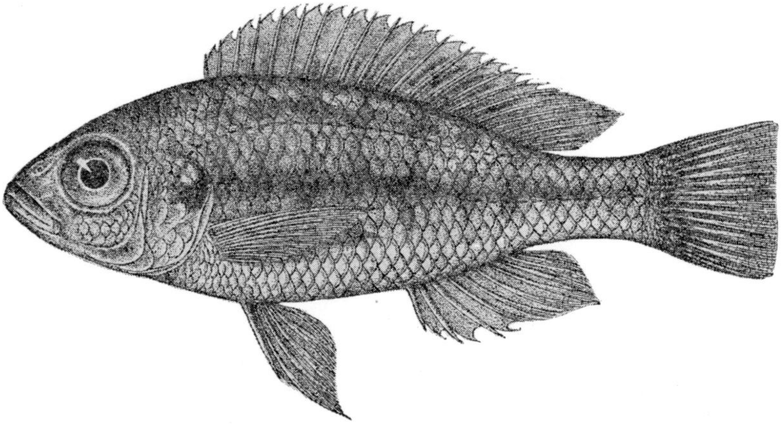

FIG. 8. *Haplochromis macrops*, an insectivore. (Natural size.)

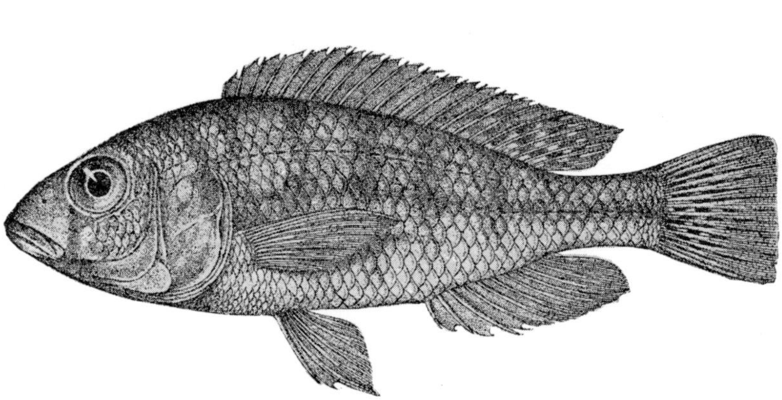

FIG. 9. *Haplochromis pallidus*, an insectivore. (Natural size.)

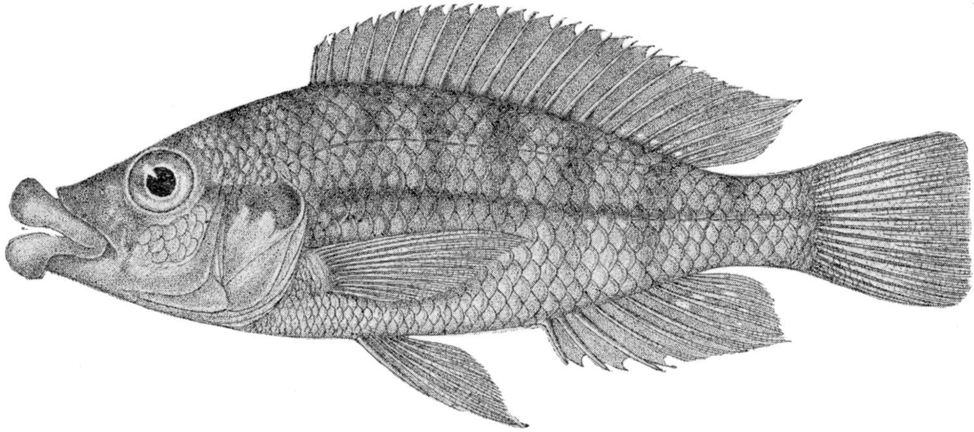

FIG. 10. *Haplochromis chilotes*, a specialized insectivore ; see p. 31. (Natural size.)

these species, strongly indicated for the seventh and assumed on anatomical grounds, for the eighth species.

Apparently none of these paedophagous species preys on the embryos or larvae of non-cichlid fishes. This is not surprising when one recalls that most non-cichlids breed outside the lake (Greenwood, 1966 ; Corbet, 1960) and then only biannually. Cichlids, on the other hand, spawn throughout the year (see p. 51). However, all the Lake Victoria *Haplochromis* (and other cichlids) for which data are available brood the embryos and larvae in the female parent's mouth ; there are no indications whatsoever that any cichlid in the lake does not practise mouth-brooding. One might reasonably conclude, therefore, that the paedophages have evolved some method of obtaining their food from the mouth of a brooding female (but see below).

Two lineages are involved in this trophic specialization (see p. 75), with both showing a certain degree of convergence (*pace* Fryer & Iles, 1972 : 101). For instance, there is a marked trend towards increased distensibility and protrusibility of the mouth, and a tendency for the teeth to be buried in a thickened oral mucosa (so much so that superficially the more specialized species in each lineage appear to be edentulous). The least specialized species in each lineage, *H. cronus* (Text-fig. 16) and *H. barbarae* (Text-fig. 29), differ little from the trophically generalized species at their particular level of overall specialization (see Greenwood, 1959b and 1967 for the species respectively). That is, *H. cronus* resembles certain species in the benthic omnivore-insectivore grade, and *H. barbarae* species in the near basal piscivore grade.

Precisely how the paedophagous predators obtain their food has not been established. The turbid water of Lake Victoria effectively prevents underwater observations, and aquarium studies have yielded equivocal results. In an aquarium *H. parvidens* (Text-fig. 18) did show behavioural patterns suggesting that it might engulf the snout of a brooding female, and in this way force it to disgorge its brood. Certainly the very distensible mouth characterizing all but the two most generalized paedophages lends support to the idea of snout-engulfing. The deeply embedded teeth also give support to this idea, since hidden in this way there would be less chance of the attacker becoming too firmly attached to the brooding female. If the predator has to struggle to free itself there is a danger of it losing its meal in the process.

On the other hand, a greatly distensible mouth could be an adaptation for sucking in the small, dispersing prey should the mother be induced by some other means to jettison her brood. In either eventuality it must be emphasized that brooding female *Haplochromis* are extremely tenacious of their broods, even when under great stress, like capture in a seine-net, handling after capture, or being chased around an aquarium (personal observations).

Fryer & Iles (1972) do not think it likely that paedophagous species get their food by directly attacking or threatening the parent fish. Instead, they believe that the predators rely on brooding females jettisoning a brood without direct interference from external sources. That females should voluntarily abandon their broods is thought by Fryer & Iles to be one of several homeostatic population density control mechanisms operating within the flock.

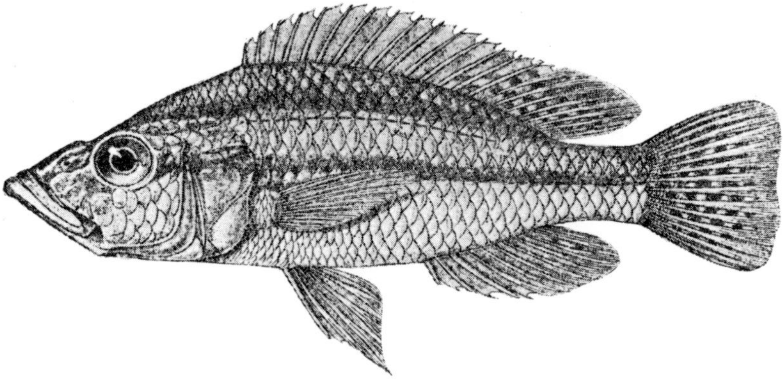

FIG. 11. *Haplochromis serranus*, a piscivore. (About $\frac{2}{3}$ natural size.)

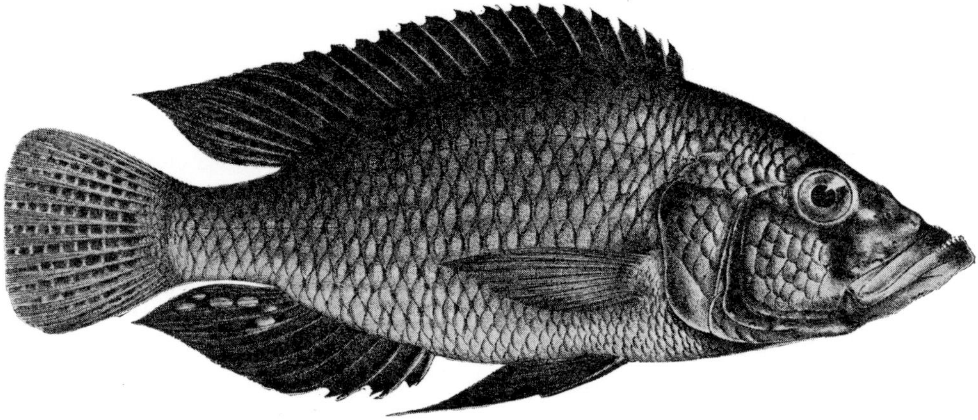

FIG. 12. *Haplochromis spekii*, a piscivore. (Half natural size.)

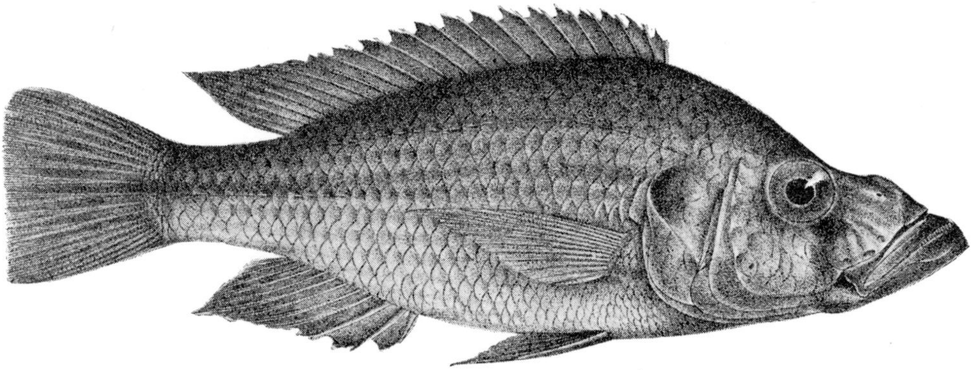

FIG. 13. *Haplochromis prognathus*, a piscivore. (Half natural size.)

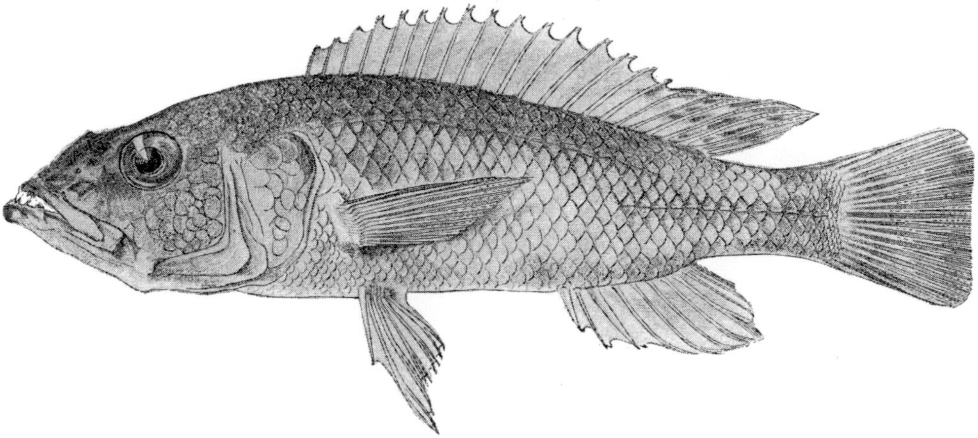

FIG. 14. *Haplochromis estor,* a piscivore. (Half natural size.)

FIG. 15. *Haplochromis mento,* a piscivore. (Half natural size.)

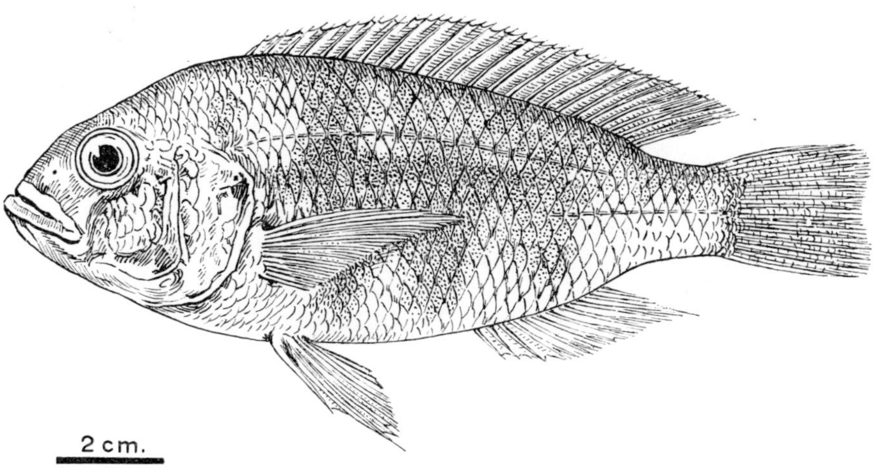

FIG. 16. *Haplochromis cronus,* a paedophage ; see p. 33.

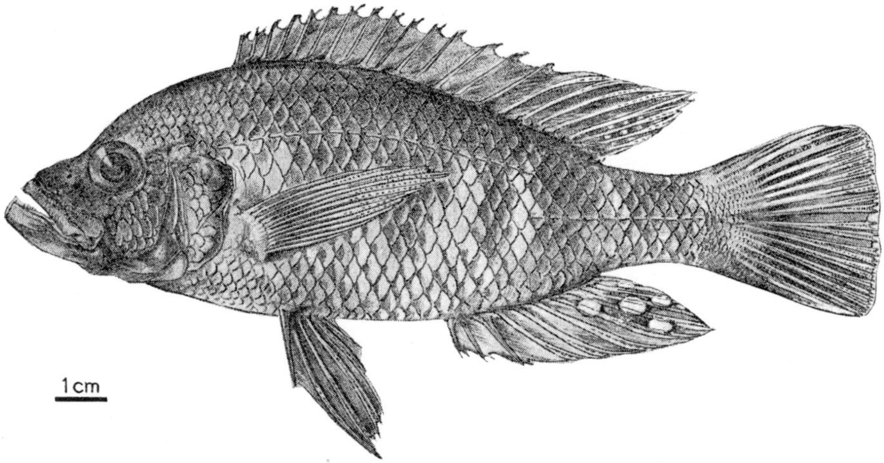

FIG. 17. *Haplochromis maxillaris*, a paedophage.

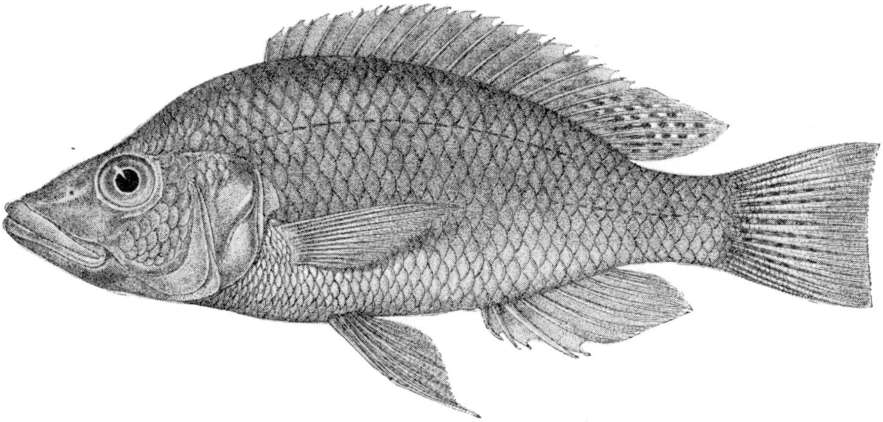

FIG. 18. *Haplochromis parvidens*, a paedophage. (Half natural size.)

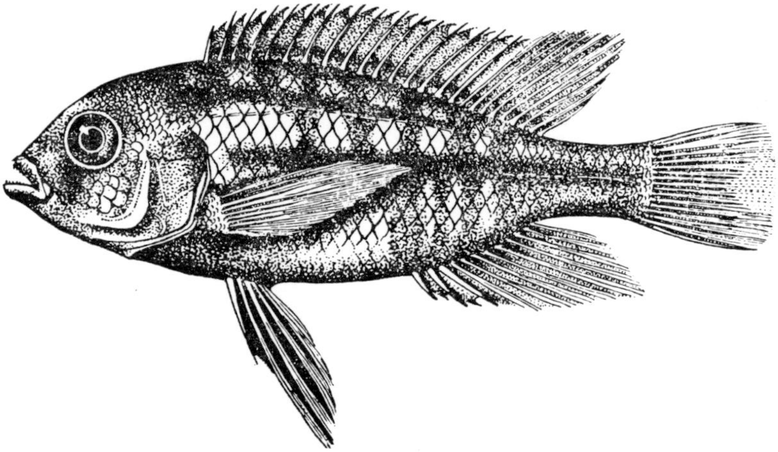

FIG. 19. *Haplochromis sauvagei*, a mollusc eater (oral sheller, see p. 69). (Natural size.)

With no direct evidence to support either suggested mode of food gathering, further speculation is pointless. However, I cannot agree with some of the evidence (or interpretation of observations) that Fryer & Iles (*op. cit.* : 101) use in support of their thesis for homeostatically induced jettisoning of the brood. As this evidence is concerned mainly with anatomical features and questions of physiology, discussion is deferred to a later page (p. 75 *et seq.*).

As to the origin of paedophagy, I would agree with Fryer & Iles (*op. cit.*) that it, like other trophic specializations, stemmed from an originally facultative habit later selected for, and further developed by, the evolution of more specific anatomical specializations. Certainly the existing 'basic' paedophages, *H. cronus* and *H. barbarae*, could not be excluded from other trophic groups on anatomical grounds alone.

Haplochromis species feeding on Mollusca form a fairly well-defined group which does, however, intergrade with the insectivorous one ; some species have a mixed insect and mollusc diet.

Ten species can be classed as predominantly molluscivorous. In this group there is a clear-cut dichotomy into species that crush the prey in the pharyngeal mill (six species) and those either wrenching the snail from its shell or crushing the shell between the jaws before swallowing the soft parts (four species). It will be recalled that at least two monotypic genera (*Hoplotilapia* and *Macropleurodus*) belong to the second group, a third (*Platytaeniodus*) may do so, and that a fourth (*Astatoreochromis*) belongs to the first category.

Those species crushing or dismembering molluscs orally form a distinct phyletic assemblage (Greenwood, 1957 ; and p. 69 *et seq.* below) which includes one of the monotypic genera, *Macropleurodus bicolor*. Snails are the chief food organisms of this group.

The species that crush molluscs (both bivalves and gastropods) intrapharyngeally are of diverse relationships (see p. 72). A common feature in all, however, is some degree of hypertrophy in the pharyngeal bones, particularly the lower one, and some molarization of the dentition on these bones (again, particularly the lower one, see Text-fig. 5). In the virtually exclusive mollusc eaters (*H. ishmaeli* and *H. pharyngomylus*, Text figs 3 and 23), these bones and their dentition reach the peak of development seen in this trophic group (see Text-fig. 5 and Greenwood, 1960). Other members of the group (e.g. *H. theliodon*, *H. humilior* [Text-fig. 22], *H. riponianus* [Text-fig. 42] and *H. obtusidens*) have the lower pharyngeal bone but slightly stouter than that in generalized species ; usually only the median tooth rows show any molarization. It is these species which, to a certain extent, are the mixed mollusc-insect eaters, and in which bivalves predominate over gastropods in the diet (Greenwood, 1960).

In terms of a restricted diet, and in associated anatomical specializations, the mollusc-eating *Haplochromis* (particularly the pharyngeal crushers) clearly show species at every major stage in the development of the trend (Text-fig. 5A, B, C). It seems likely that the anatomical ability to crush mollusc shells has evolved, independently, on a number of occasions. But at least one phyletic line can be detected, with representatives of its insect–mollusc eating species, bivalve–gastropod eating species and almost exclusively gastropod eating species, still extant (see pp. 69–75 below).

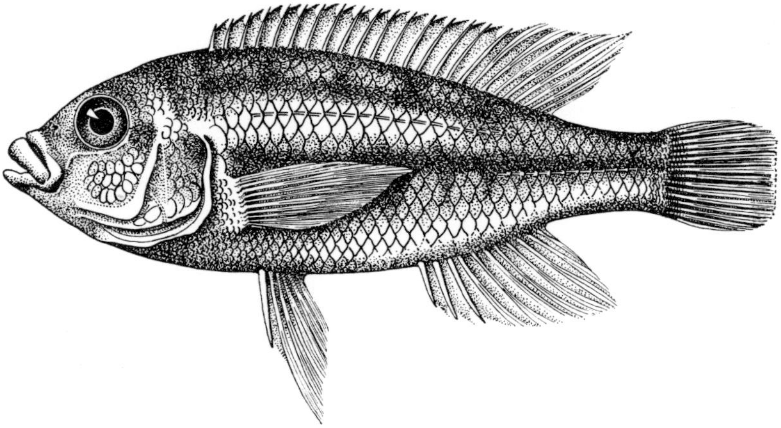

FIG. 20. *Haplochromis granti*, a mollusc eater (oral sheller). (About natural size.)

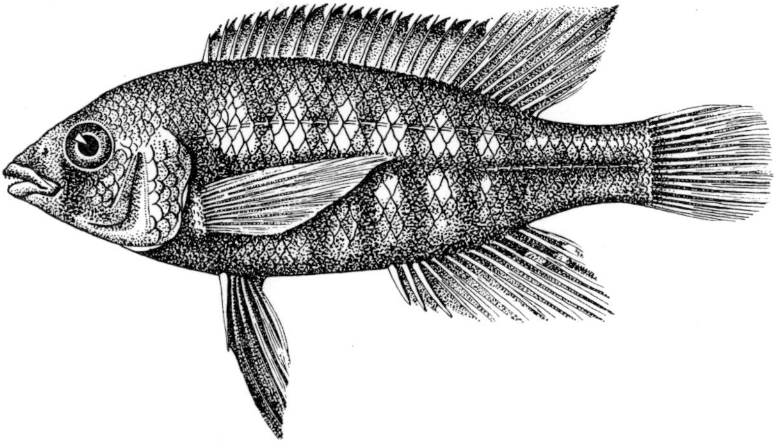

FIG. 21. *Haplochromis xenognathus*, a mollusc eater (oral sheller). (Natural size.)

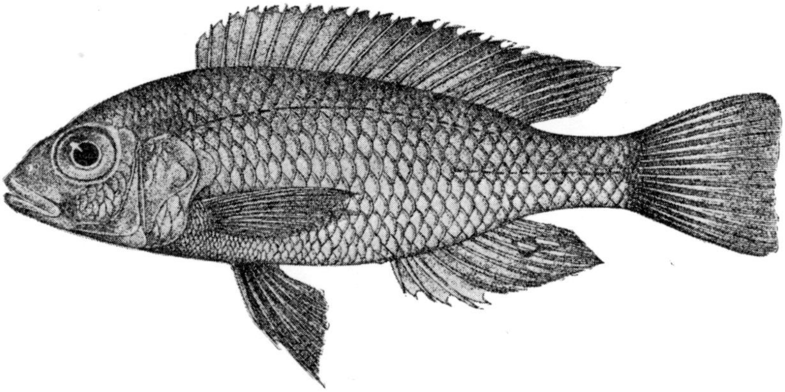

FIG. 22. *Haplochromis humilior*, a mixed mollusc–insect eater, the molluscs are crushed in the pharynx. (Natural size.)

Small Crustacea, especially ostracods and, to a lesser extent, prawns of the genus *Caridina*, contribute to the diet of several *Haplochromis* species in the insectivore and the mollusc–insectivorous trophic groups.

There is, however, a small group of five species that appear to feed equally on insects and *Caridina* (Greenwood & Gee, 1969; Greenwood, 1967 for *H. tridens*). All are bottom feeders and most occur in water 20–35 m deep (with one species extending into shallow water). Almost certainly more species will be added to this group when undescribed material from the deep waters of the lake is analysed. No obviously adaptive anatomical features are associated with the insect–crustacean diet; all the species in this group, with one possible exception, appear to be closely related (Greenwood & Gee, 1969; and p. 67 below; Text-figs 24 and 25).

Plant material (especially algae) is frequently found in the guts of bottom feeders whose food is otherwise of animal origin. The nutritive value of the plant material in the diet of such species is unknown; usually the quantity ingested is small and rarely does it appear to be digested completely.

Eight species, however, show the characteristically long, much coiled intestine and the highly distensible stomach, associated with a truly vegetarian diet. The gut contents of all eight species confirm their placement in this trophic category. Once again, there are indications from the unstudied material at my disposal that the number of phytophagous species will be increased eventually.

The plant eating species can be split into three subgroups: those feeding on epiphytic and epilithic algae, those feeding on phytoplankton and those eating macrophytic plants.

Since only one species (*H. phytophagus*; Greenwood, 1966b) is known from the latter subgroup it can be discussed briefly. *Haplochromis phytophagus* has the overall structural and dental characters of a generalized species, except that the teeth (both oral and pharyngeal) seem adapted for biting off and then macerating the relatively tough tissues of higher plants. Analyses of gut contents suggest, however, that the principal source of food is not this tissue itself, but the associated flora of attached diatoms. Only the ruptured leaf and stem cells appear to be digested.

The four species feeding directly on epilithic and epiphytic algae (Greenwood, 1956b) obtain their food in a different way from that of *H. phytophagus*. They graze the algae from the substrate and little higher plant tissue is ingested (and, as in *H. phytophagus*, this is poorly digested). One of the four species (*H. nigricans*; Greenwood, *op. cit.*) is predominantly a grazer on epilithic algae and is rarely caught away from rocks and rocky shore lines.

The other three species (*H. obliquidens*, *H. lividus* and *H. nuchisquamulatus*) sometimes extend into the habitat range of *H. nigricans* and also graze off rocks. Usually, however, these species are found in or near stands of rooted plants.

All four species feeding on periphyton show some dental modifications (Text-figs 4D–G and 36), with *H. obliquidens* the most obviously specialized in this respect (Text-fig. 4G). The phyletic–structural story here is interesting because, despite the different degrees of specialization in tooth shape (from near generalized in *H. nuchisquamulatus*, through *H. lividus* to highly specialized in *H. obliquidens*), there is apparently little difference in their effectiveness as scrapers. The teeth of *H.*

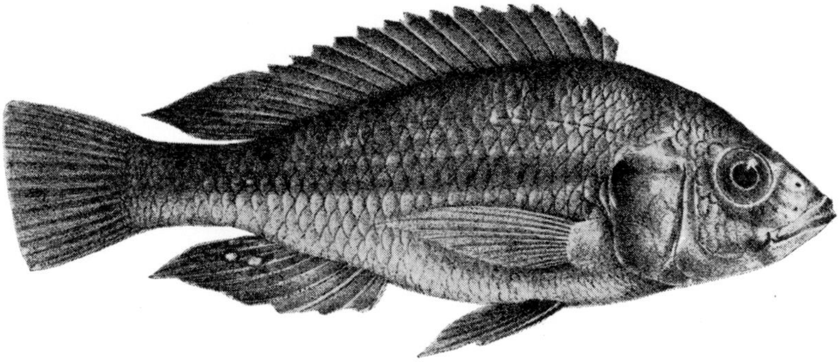

FIG. 23. *Haplochromis ishmaeli,* a mollusc eater (pharyngeal crusher). (Natural size.)

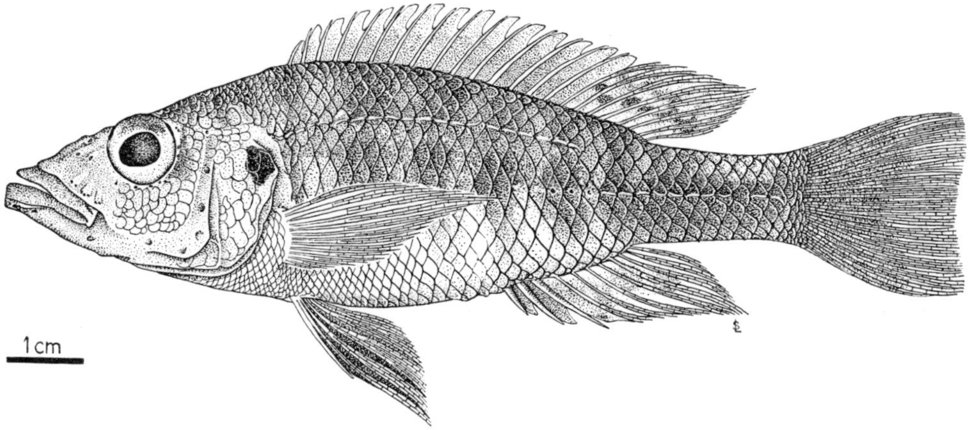

FIG. 24. *Haplochromis dolichorhynchus,* a benthic crustacean eater.

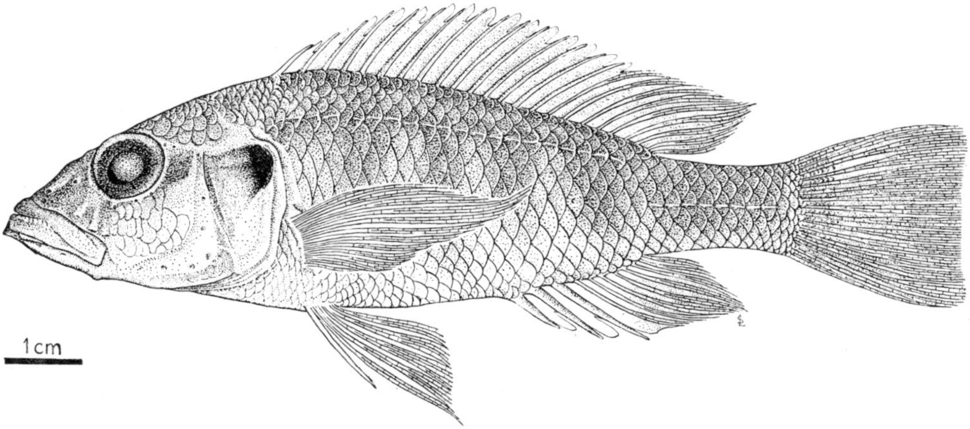

FIG. 25. *Haplochromis tyrianthinus,* a benthic crustacean eater.

nigricans (Text-fig. 36) are quite different, and probably are associated with its diet of epilithic periphyton.

The three phytoplankton feeders (*H. erythrocephalus* [Text-fig. 26], *H. paropius* and *H. cinctus* [Text-fig. 27]) are in all respects, except the long gut, typical members of the structurally generalized species groups. None shows any obvious superficial characters correlated with feeding on finely particulate matter. The gill rakers, for example, are neither longer nor more closely arranged than in an insectivore (Text-fig. 6), although in *H. erythrocephalus* there is a slight increase in number (12 or 13 *cf* 8–10 for the other phytoplankton feeders and most insectivores).

Judging from the nature of the algal and other material found in the stomachs of these fishes, food is taken from the bottom and not while it is in suspension (Greenwood & Gee, 1969).

Finally, there is the enigmatic *H. acidens* (Text-fig. 28), an apparent phytophage with a distinctly predatory facies. The gut contents of the 16 fishes examined (from different localities) consist mainly of finely macerated phanerogam tissue, but also a few insect remains, some bones of small fishes and in one specimen the soft parts of a mollusc. The intestine of *H. acidens* is proportionately somewhat longer than that in omnivores, but is shorter than in other vegetarian species. To confuse the issue further, the jaws, skull and teeth of this species are typically those of a piscivorous predator (see Text-fig. 37 ; and Greenwood, 1967 ; also p. 67 below). It was with a piscivorous lineage that I originally suggested *H. acidens* be related (Greenwood, *op. cit.*).

Fryer & Iles (1972) call *H. acidens* an '. . . adventurous species, formerly piscivorous, which made experimental sorties into plant eating . . . an almost vacant niche in Lake Victoria'. They go on to note that such a trophic shift would be one from a way of life in which specialization was at a premium to one in which the reverse is true, thus explaining what might be considered 'imperfections' for phytophagous habits in the jaws and teeth of *H. acidens*. This idea is intriguing but it must be remembered that there are trophically generalized species not far removed anatomically from the basal species in the piscivore radiation with which *H. acidens* was first associated (but see p. 67 below). Recent studies on this species, however, indicate that it could as well be related to a generalized and omnivorous species in a lineage that produced no predators but which did produce another herbivore (*H. erythrocephalus*). The 'predatory facies' of *H. acidens* is thus all the more difficult to explain.

The last trophic group to be considered comprises a single species, *H. welcommei* (Text-figs 3 and 39), the only known lepidophagous species in the lake. The guts of all specimens examined, contained numerous small fish scales (like those covering the caudal fin base of *Haplochromis*), together with a few fragmentary fin rays and bits of skin. (The number of guts checked is now considerably greater than when the species was first described [Greenwood, 1966b], but is still too small for satisfactory generalizations to be drawn.)

Scale eating is practised by many cichlid species in Lakes Malawi and Tanganyika (reviews in Greenwood, 1966b and Fryer & Iles, 1972). Some of these species, like *H. welcommei*, scrape the caudal fin squamation from other cichlids ; others nip

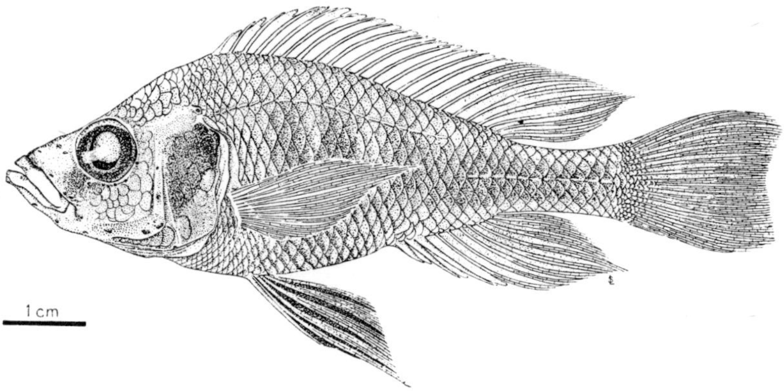

FIG. 26. *Haplochromis erythrocephalus*, a phytoplankton eater.

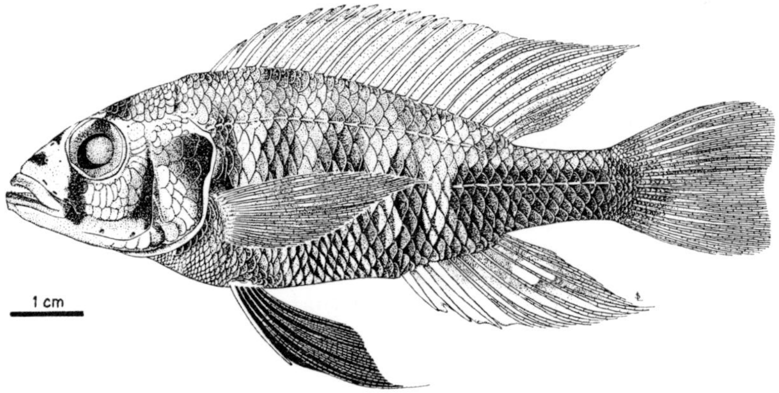

FIG. 27. *Haplochromis cinctus*, a phytoplankton eater.

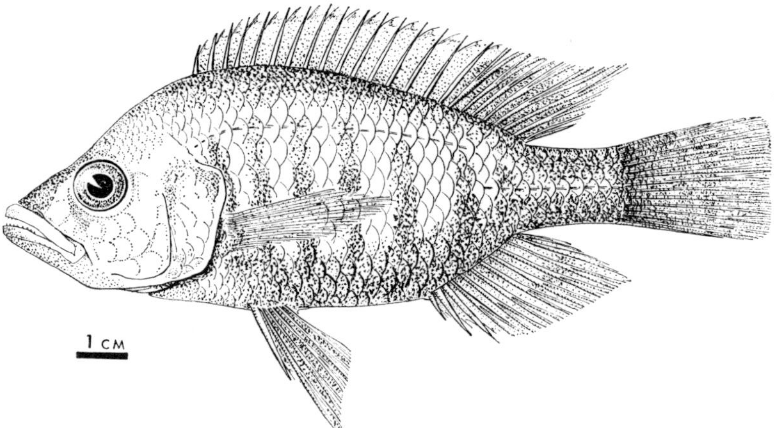

FIG. 28. *Haplochromis acidens*, probably a feeder on macrophytes, see p. 41.

scales from the body of a variety of fishes. The diet of *H. welcommei* is, however, not entirely one of scales and fin rays ; most guts also contain a quantity of the diatom *Melosira*, and in some there are fragments of macerated plant epidermis. This plant material suggests that *H. welcommei* may also feed by grazing on the epiphytic flora of rooted plants, for which habit its dentition would seem well adapted (Text-fig. 39 ; also p. 69 below). If this is so, then the species shares a trophic niche with *H. obliquidens* and the other grazers (see above, p. 39).

Interestingly, one of the Lake Malawi lepidophages, *Genyochromis mento*, also feeds on periphyton (Fryer, 1959) which it browses rather than grazes from rock surfaces. Fryer (*op. cit.*) suggests that *G. mento* could have evolved from the same specialized stem as some of the epilithic algal grazers in that lake. *Haplochromis welcommei*, on the other hand, does not seem to be at all closely related to the grazing species of Lake Victoria. Its relationships are apparently with the benthic, crustacean-eating species discussed on p. 67 (see also Greenwood, 1966b).

There are still several of the described *Haplochromis* species whose feeding habits are unknown, or for which so few and contradictory data are available that the species cannot be put in any particular trophic category. Furthermore, there are at least thirty undescribed species collected recently from several deepwater habitats in the centre lake region. Preliminary studies indicate that most belong to the insectivore and crustacean-eating groups, with two or three other species referable to the piscivores.

The absence of a pelagic zooplankton eating *Haplochromis* species has already been commented upon. No reason for this obvious gap in the trophic radiation is immediately apparent. There is, of course, a non-cichlid *Engraulicypris argenteus* occupying this niche, but then there are non-cichlid species in all the other trophic niches occupied by *Haplochromis* species living in the same habitat. Possibly there is a heavy predation of zooplankton by young fishes (and other animals) in the littoral and sublittoral lake regions, so that the niche is, in fact, fully exploited. This purely speculative suggestion would not seem to apply to the offshore, open-water areas of the lake (except for at least their partial occupancy by shoals of *E. argenteus* ; see Graham, 1929). Here again one is hampered by a lack of information on zooplankton density and distribution, and by a complete ignorance of what trophic categories of cichlids there are in the pelagic and midwater zones of the lake.

Before moving on to consider other aspects of *Haplochromis* biology one should recall that the adaptive radiation described above has taken place in a *single* genus. To put this phenomenon in perspective one should also remember that in any major tropical marine biotope (say a coral reef) there might be as much trophic diversity among its fishes, but several phyletically distinct families would be involved (Greenwood, 1965a).

Particularly instructive in this respect, since it involves a lacustrine community (albeit a cold temperate one), is the work by Keast & Webb (1966) on the feeding habits of fishes in Lake Opinicon, Ontario. These authors analysed the mouth and body forms, and the feeding habits of 14 species belonging to seven families (representing, incidentally, five orders). In terms of their overall function and gross morphology (except fin position) the characteristics of these different taxa are all apparent in various Lake Victoria *Haplochromis* species. What is more, the *Haplochromis* have exploited more trophic niches than have the fishes of Lake Opinicon.

Keast & Webb (*op. cit.*) conclude that mouth and body structure combine with food specializations and habitat preferences to reduce greatly interspecific competition. By contrast, in Lake Victoria it is usual for more than one *Haplochromis* species to occupy the same trophic niche in the same habitat. There is, however, a similarity in the two lake situations, for the *Haplochromis* species, like the fishes of Lake Opinicon, retain a measure of flexibility in their feeding habits (see data in papers by Greenwood, 1956–69). Keast & Webb (*op. cit.* : 1871) consider such flexibility an important feature for survival in cold temperature lakes ; it may well be equally important in warm tropical lakes, and have played a part in the evolution of the *Haplochromis* species flock (see p. 115 below).

The Lake Victoria *Haplochromis* are, of course, not unique in these respects. The Lake Malawi species are even more adaptively multiradiate, and occupy some niches not even represented among the Victoria flock (although paedophagy seems to be a unique feature of that flock). For a detailed account of the situation in Lake Malawi the reader is referred to the excellent account given by Fryer & Iles (1972).

Intralacustrine distribution of the *Haplochromis* species

When considering the distribution, habitats and vertical range (i.e. depth distribution) of the Lake Victoria *Haplochromis* species the same restrictions must be applied as were applied to data on feeding habits (see p. 30 above). Furthermore, the murky waters of the lake (transparency 1·3–8·2 m, depending on habitat, lowest in sheltered bays) have greatly impeded field observations on the fishes. Thus, information on the distribution of a species within a major habitat has to be got indirectly, and its precision is largely dependent on the type of collecting gear used. For example, a trawl or seine net could pass through clumps of plants, could traverse several types of bottom and could take fishes at a variety of depths. Lack of rigorously quantifiable data on the number of individuals in any one habitat is also a hindrance in assessing species' distributions and in determining habitat preferences.

Major habitat types within the lake tend to intergrade with one another, but are recognizable as :

(i) Exposed beaches with some wave action. The substrate here is generally sandy, with small rocks and rocky outcrops ; some rooted plants occur, and there are often marginal stands of emergent vegetation. The sandy substrate extends offshore for variable distances, grading gently into the organic mud substrate characteristic of sheltered bays and the deeper parts of the lake.

(ii) Sheltered bays, usually fringed with papyrus swamps ; often the inshore end of a bay is an extensive papyrus swamp. In most sheltered bays the bottom is composed of organic mud extending right up to the marginal papyrus ; sometimes, however, there are narrow sand beaches, particularly if the papyrus margin is interrupted.

(iii) and (iv) Sub-littoral regions (water depth 6–20 m), and deepwater regions (more than 20 m deep). These are rather more artificial than natural categories, and are used in order to divide the offshore zones into 'shallower' and 'deeper' waters. The sublittoral substrate is either organic mud, shingle, or mud with low outcrops of lateritic rock that extend for several metres as flat sheets rising one or two metres above the prevailing substrate. This zone grades imperceptibly into the 'deepwater' one where the substrata are also similar.

Islands of various sizes (from Buvuma island, about the size of the Isle of Wight, to mere rock pinnacles, *ca* 10–20 m across) are common in the sublittoral zone.

Depending on the island's size, its shoreline can be as varied as that of the mainland. Larger islands also provide the chief areas of truly rocky habitats.

(v) Coastal littoral other than sandy beaches. Apart from a few regions of the lake where there is a rocky coastline, much of this littoral region is adjacent to fringing papyrus and often includes a water-lily swamp zone. The bottom can be hard (sand or shingle) or soft (organic mud). The papyrus swamps (even narrow fringing ones) rarely harbour cichlids, except at the perimeter where, unlike water further into the swamp, the dissolved oxygen level is adequate for these fishes. Otherwise the chief occupants of such swamps are air-breathing non-cichlids (see p. 11).

The large gulfs and sounds (e.g. Kavirondo Gulf, Napoleon Gulf, Smith Sound) are not in themselves particular habitats ; each contains most of the habitats detailed above.

An indication of the number of *Haplochromis* species occupying a particular habitat can be derived from an exposed beach near Jinja (Nasu Point beach) and a sheltered, papyrus fringed bay about 8 km south of Jinja. In both places I was able to carry out regular sampling over a period of six years.

The beach habitat is occupied by 35 *Haplochromis* species, the bay by 21 species. In the former habitat few non-cichlids (*Barbus* spp. and *Clarias mossambicus*) appear to be regular inhabitants, and then only in very small numbers. Occasionally, however, the beach was invaded by large shoals of the characin *Alestes sadleri*, or by shoals of juvenile *Tilapia variabilis* (otherwise not a regular component of the cichlid faunas in that habitat).

Ekunu Bay, by contrast, carries a varied non-cichlid fauna including *Protopterus aethiopicus*, *Synodontis victoriae*, *Clarias mossambicus*, *Bagrus docmac*, *Mormyrus kannume*, *Gnathonemus* spp., *Labeo victorianus* and, occasionally, the large *Barbus altianalis*. Both *Tilapia esculenta* and *T. variabilis* are resident inhabitants.

The two habitats differ quite markedly in the species of *Haplochromis* present ; probably not more than two species are shared, apart from the ubiquitous *Astatoreochromis alluaudi*. There are differences too in the trophic specializations shown by the *Haplochromis* of the two habitats. In this respect Nasu Point beach is somewhat more varied, but there is a higher proportion of piscivorous predators in Ekunu Bay (13 of 21 *Haplochromis* species, *cf* 13 of 35 at Nasu Point).

Surveys made in other beach and bay habitats seem to confirm the impression of greater diversity of *Haplochromis* species from beaches, and the virtual exclusion of non-cichlids from this habitat.

Some idea of population density in the *Haplochromis* species occupying a beach and its sandy sublittoral region can be gained from the commercial catches made in these areas. Over the period 1951–58 the average annual catch from seine nets operated at a beach near Majita (Tanzania) was 2·5 million individuals ; the highest annual seine net catch ever recorded, also from a beach in Tanzanian waters (at Nyamwikumulu in 1957), was 7 589 599 individuals in a total of 317 hauls (figures from the Lake Victoria Fisheries Service Annual Reports). Both these figures include the monotypic genera (which were not distinguished from *Haplochromis* by the counters), but these species contribute only a small fraction of the whole (personal observations on these beaches). For comparison, the catch of non-cichlid fishes in the same number of seine net hauls at the same beach, Nyamwikumulu, in 1957 was 10 974.

As far as I can detect, the nature of the substrate is the major factor influencing the distribution of many *Haplochromis* species, at least within the depth limits of the various species (see below). Some species are virtually confined to habitats (or parts of a habitat) where the bottom is muddy. Others are restricted to places where the substrate is hard and composed of sand, rock or shingle. Unfortunately, the data available are not sufficiently refined to be able to tell if the exact nature of the substrate, that is its constitution, or its associated flora and fauna, has the more particular influence on species distribution.

Of the 69 species apparently restricted to water less than 20 m deep (i.e. inshore species), 27 are found only over hard substrates, and five only over organic mud or other soft substrates. The remaining 37 species have been caught over both hard and soft bottoms, but it must be stressed that of these species, nine rarely occur over soft substrates and seven are rare over hard bottoms. In other words, 32 of the 69 species have a restricted substrate preference, and 21 occur over either substrate type with sufficient frequency for one to consider them as being free from substrate restraint (data from Greenwood, 1956–69).

Correlating specific substrate preferences with the trophic groups to which the species belong is difficult, and because of insufficient data also rather subjective and imprecise. Some species in all trophic groups, except the algal grazers, show no marked substrate preferences. Molluscivorous species of the group crushing gastropods orally (or wrenching them from the shell) are virtually confined to hard substrata, as are the paedophagous species (see pp. 31–37 above). All other trophic types are found over every type of substrate.

Likewise it seems that representatives of all trophic categories are found in each of the major habitats. However, the algal grazers and the species feeding on higher plants (p. 39 above) are restricted to habitats or parts of a habitat where there are rooted plants or, in the case of the grazers, suitable substrata for algal growth. *Haplochromis nigricans* of this group seems to have the most restricted distribution, rarely being found far from rock surfaces. The paedophages as a group are relatively restricted to littoral habitats, probably because it is mainly in these regions that brooding female *Haplochromis* are concentrated. Particular interest attaches to six specimens of *H. obesus* (or a taxon anatomically very like that species) caught over a mud bottom at a depth of 27–30 m (Greenwood & Gee, 1969). Collections of *Haplochromis* from this and similar habitats rarely produce brooding females (or individuals showing the characteristic buccal distortion that goes with brooding and which persists for a short while after the young have been jettisoned). Thus it is especially noteworthy that these deep-living *H. obesus* had all fed exclusively on cladocerans and copepods.

It is clear from field observations that more than one *Haplochromis* species of a particular trophic group can be found in any one habitat. The extent of such interspecific overlap, and hence possible trophic competition, may be less than is implied by the mere record of several species seeming to feed on the same food source.

Two algal grazers, *H. lividus* and *H. obliquidens*, occur together in the sandy beach habitat ; but the former species extends its range into deeper water than does *H. obliquidens*. *Haplochromis nigricans*, with similar food requirements, also occurs in this habitat but it feeds principally from submerged rocks and stones.

A similar situation exists with two mollusc eaters in this habitat, *H. sauvagei* and *H. prodromus*, where *H. prodromus* extends into the sublittoral beach zone. Furthermore, the two species have different feeding methods, *H. sauvagei* levering the snail from its shell, and *H. prodromus* crushing the shell between its jaws. These differences may influence the selection of prey species by the predators. The monotypic genus *Macropleurodus bicolor* also enters the picture since its habitat range overlaps in part those of *H. sauvagei* and *H. prodromus*, being very like that of the latter species (Greenwood, 1956a, 1957). Its feeding habits, too, are more like those of *H. prodromus*, although it is known to feed on insects as well as snails (Greenwood, *op. cit.*), and probably to a greater extent than *H. sauvagei*, also a mixed feeder.

The whole situation is further complicated by the presence of two other mollusc eaters of the same trophic complex, *H. granti* and *H. xenognathus* (see p. 69 *et seq.*). Both these species occur in the habitat under discussion, and both share the same food sources and intrahabitat preferences as the species already considered. Nor are these the only mollusc eaters present. *Haplochromis humilior* and *H. pharyngomylus* also live in beach habitats (Greenwood, 1960). These two species, unlike the others, crush their prey in the pharyngeal mill ; the food of *H. pharyngomylus* (chiefly the snail *Melanoides tuberculata*) is virtually identical with that of *H. prodromus*, *H. xenognathus*, *H. granti* and *M. bicolor*. The food of *H. humilior*, a smaller fish, overlaps that of these species but does include a greater proportion of other organisms, especially insects (Greenwood, 1960). Other partly molluscivorous species could be mentioned but I think the problem is sufficiently obvious without introducing additional complexities.

An equally complicated picture can be compiled for the insectivorous species in a habitat, and to a lesser extent for the phytoplankton eating species as well.

The piscivorous predators provide perhaps the strongest *prima facie* case for interspecific trophic congruence ; yet even here interspecific competition cannot definitely be established from the data available. In the habitats I have studied there seems always to be a group of piscivorous species with completely overlapping ranges, as well as a few species showing a restricted distribution. All these fishes feed on other *Haplochromis* species. Since it is impossible to identify prey recovered from the guts to more than the generic level, it is equally impossible to tell what degree of competition (if any) there is for prey species. On a subjective impression there would seem to be a sufficiency of prey species for competition between the predator species not to develop, and the situation is further ameliorated by some of these having a limited intrahabitat distribution.

The extent to which some species may move between habitats, or from one type of substrate to another, has not been established. Thus I would be chary of suggesting, as have Fryer & Iles (1972 : 308), albeit tentatively, that in Lake Victoria '. . . It seems probable that some species will at times utilize this ability, (to shift habitat), at least as a temporary expedient for avoiding adverse conditions'.

I would, however, agree with the general conclusion reached by these authors, that in Lake Victoria there is, as compared to Lake Malawi, less obvious stenotopy amongst members of the cichlid species flock (always bearing in mind that far less is known about the generality of *Haplochromis* species in Lake Malawi than about the 'Mbuna' generic complex and the few rock-haunting *Haplochromis* species that coexist with them, a point not explicitly made by Fryer & Iles).

None of the Lake Victoria *Haplochromis* or related species shows any geographical restriction within the lake, nor have any morphologically distinguishable populations been discovered. In both these respects the Victoria species differ from those of Lake Malawi, and especially Lake Tanganyika (see summary in Fryer & Iles, 1972).

The question of interspecific overlap in feeding habits and habitat requirements among Lake Victoria *Haplochromis* species has direct bearing on the principle of 'competitive exclusion' (the so-called Gause's principle, but see Mayr, 1963). This postulated that no two species can exist at the same locality if they have identical ecological requirements. As Wynne-Edwards (1962) has demonstrated, the principle is logically and ecologically unsound. If two species have truly identical requirements they will be, in effect, ecologically one species. Competition will then be on an intraspecific rather than an interspecific level ; that is, between individuals.

From what is known about the inshore *Haplochromis* species considered so far, there are relatively few, if any, aggregates with absolutely identical ecological requirements (i.e. a condominium, *sensu* Wynne-Edwards). Instead there are many species with a certain degree of overlap. Following the spirit of Professor Wynne-Edward's terminology, perhaps such groups should be thought of as a 'commonweal'?

The vertical movement of inshore *Haplochromis* species is very poorly known, and their apparent restriction to water less than 20 m deep may prove illusionary. Taking the latter point first, most available depth records were obtained before the deeper, offshore areas of the lake were studied intensively. In recent years there has been a great deal of exploratory trawling in these areas. To a certain extent this research has confirmed the earlier depth distribution records, but it has also provided some surprises. The extended range of *H. obesus* has already been noted (p. 46 above). During a recent trawl survey a haul made over sand at a depth of *ca* 50 m in midlake (opposite Entebbe) contained, in addition to deepwater species, several other species that were previously known only from inshore beach habitats (personal observations).

Little can be said about the movement of inshore species through the water column. The indications are that species in this zone remain near the bottom, at least during daylight (personal observations). In water less than 2 or 3 m deep, many species may range from surface to bottom, but in progressively deeper water their behaviour pattern becomes increasingly benthonic. Judging from the presence of bottom debris associated with food in the guts, most species feed on or near the bottom. For the piscivorous species, such indirect evidence is not available; the occasional field observation in clear water suggest that piscivores range rather freely through the water column, at least in shallow parts of the lake.

As yet the taxonomy of deepwater species is very incompletely known, and the ecology of even the described species is still imperfectly documented (Greenwood & Gee, 1969). Apart from a few inshore species whose range extends beyond *ca* 20 m, 17 described species are known from deeper waters (Greenwood & Gee, *op. cit.*; Greenwood, 1967). I estimate that at least 30 more species will be described from collections already obtained. Only two species, *H. dolichorhynchus* and *H. erythrocephalus*, have been recorded from depths less than 15 m; these species have depth ranges of 10–30 and 10–35 m respectively. Depth ranges for the remaining species are between 17 and 70 m. Most have a wide range within the limits of 25–35 m, but there are indications of some species being restricted to depths near the upper and lower limits of the total range. None extends into the truly littoral zone, but several species may live at depths greater than those recorded here (personal observations).

Deepwater species are necessarily confined to the strictly offshore regions of the lake, but at least three species (*H. erythrocephalus*, *H. dolichorynchus* and *H. cryptogramma*) have been caught in the deeper areas of large bays (Greenwood & Gee, 1969).

It is impossible to define habitats in the deeper water, and the collecting gear employed there (trawl nets) means that several habitats may be sampled in one trawling period (half an hour at a speed of about 2 knots). That all but two of the known species were caught over a soft bottom is probably not a true reflection of the species' substrate preferences; more trawling has been done over mud than over hard substrata. Nor, because of limits imposed by the collecting methods, is it possible to determine the species structure of any particular area. All one can say is that several species are caught in a trawl, that these species will be from several trophic groups, and that any particular trophic group is usually represented by more than one species. In other words, a replication of the situation in shallower waters of the littoral and sublittoral.

The structure of deepwater communities may, however, be simpler because certain trophic groups are not represented there. So far, insectivores, bottom detritus feeders, predators on small Crustacea and piscivorous predators have been identified (Greenwood & Gee, 1969). No mollusc eaters have been found, and the sole paedophage (or presumed paedophage, *H. obesus*, see p. 46 above) had fed on Crustacea. These remarks must, however, be qualified by reference to the 'inshore' species complex caught in midlake (see p. 48); at least two mollusc eaters (including *Hoplotilapia retrodens*) were amongst the species then caught.

The apparent paradox of that deep-living 'shallow water' community shows clearly the difficulty of generalizing about the ecology of deepwater regions in the lake. If sandy bottom, inshore species can adapt to deep water, why is it that none of the soft bottom inshore species has done likewise, especially since the nature of the soft substrata in both regions is apparently identical? Much more will have to be learned about deepwater fishes, and particularly if there are species confined to sand at those depths, before an attempt can be made to answer this question.

It is assumed from the evidence available (particularly the nature and composition of gut contents) that all deepwater species are benthic in habit (Greenwood & Gee, 1969). One species, *H. laparogramma,* is suspected, however, of feeding away from the bottom (Greenwood & Gee, *op. cit.*), and other members of the insect–crustacean eating trophic group may have similar habits, or at least be facultative in this respect because their prey are not necessarily confined to the benthos.

Echo-sounding has demonstrated the presence of fish in midwater at a variety of depths (Gee, 1968). So far it has been impossible to identify these fishes ; that they may be species of *Haplochromis* is indicated by some rather inconclusive midwater experimental trawling. (Personal communication from staff members of the E.A.F.F.R.O., 1971.)

No counts are available for the number of individual fish caught per trawl in deep waters. An estimate by eye suggests that the density of *Haplochromis* in these waters is not significantly less than from inshore habitats (see p. 46 above).

Brief mention may be made here to the habit of shoaling. In Lake Malawi several of the offshore, zooplankton-eating *Haplochromis* species (the 'Utaka' group) exhibit shoaling behaviour, as does the rock frequenting *H. kiwingi* during the zooplankton-eating phase of its life cycle (see Fryer & Iles, 1972).

The Lake Victoria *Haplochromis* species do not appear to furnish a single example of shoaling, at least during adult life. Since adult shoaling is associated with feeding on suspended, particulate food (zooplankton or phytoplankton) the absence of this behavioural trait is not altogether surprising. However, it should be remembered that direct observation on fishes in Lake Victoria is well nigh impossible, and the recognition of a shoal would be from the extreme abundance of a particular species in a catch. Thus, with trawling, when several thousand individuals are caught from over a large area, a small shoal could pass undetected in the catch.

Species of the shoaling 'Utaka' *Haplochromis* group in Lake Malawi have characteristic 'shoaling marks', in the form of spots, or a well-defined stripe, on the flanks. Few Lake Victoria species show such coloration, at least as fixed patterns (see p. 52). Of the species that do, *H. martini* and *H. michaeli* (lateral stripe) provide no evidence for shoaling, and neither species is a plankton eater. The possibility remains open for two deepwater species with marked patterns (*H. laparogramma* [a lateral stripe] and *H. cryptogramma* [an interrupted lateral stripe]). Both species are caught only in trawls, thus making an estimate of their relative abundance in a catch particularly difficult. Since *H. cryptogramma* feeds on the pupae of chironomid flies (which occur in dense aggregates), and seems to feed away from the bottom, it is the most likely suspect for shoaling behaviour.

Breeding biology of the *Haplochromis* species

To conclude this review of *Haplochromis* bionomics the rather scattered information on their breeding biology will be brought together and reviewed.

Many species are known to be female mouth brooders. There is no indication, either direct or by implication, that any species does not practise this form of parental care. For the several species where females actually carrying young have not been caught there is indirect evidence of mouth brooding; namely, spent females with the characteristic deformation of the buccal cavity (a deeply depressed hyoid arch) associated with mouth brooding, and the presence of few but large ova in the mature ovary. These ova also fail to show the modifications to the *zona radiata* which are invariably associated with substrate spawning habits in the Cichlidae (see Fryer & Iles, 1972 for review).

No records are available for the breeding activity in any particular species. However, repeated observations made at a beach near Jinja, and from other localities in northern waters of the lake, strongly suggest that at least part of a population is breeding at any one time. That is to say, for the majority of known inshore species breeding is continuous. (Unpublished personal observations.) Nothing (except indirect evidence for mouth brooding) is known about the reproductive behaviour of species from deeper waters.

Regrettably the opaque water of Lake Victoria does not often permit direct observation of the fishes, and even when this is possible it is not easy to identify the species seen. Thus the exact spawning sites for inshore species, or the mode of their spawning, is not known. On occasion I have seen males guarding simple pit nests in the clear sand bottom near stands of emergent swamp grass. If nest building (or at least the preparation of a substrate for oviposition) is a common feature for all species, the question is raised of where do species living over soft substrate spawn. The flocculent organic mud covering so much of the lake floor would seem to be a most unsuitable substrate for this purpose.

This leads one to consider the possibility of species which live over soft mud during non-breeding phases moving to areas more suitable for spawning. Brooding females of some of these species are, however, found in the same areas as non-breeding fishes.

The same question is raised with regard to the deepwater species. None has been found inshore so presumably breeding takes place offshore, possibly over those areas where the bottom is hard. There are numerous other unanswered problems associated with the breeding habits of these species. For instance, if there is a migration to hard substrata, is there competition for breeding space with species normally resident in such places, or at least conflict between breeding immigrants and feeding residents?

Then there is the question of light and vision in deep water. Visual signals and stimuli play an important part in cichlid courtship and reproductive behaviour (see extensive summary of researches in this field given by Fryer & Iles [1972], and papers by Baerends & Baerends van Roon [1950], Wickler [1966] and Neil [1964]). Light penetration in Lake Victoria is low even in water less than 10 m deep; it must be even less at greater depths. For the moment we are nowhere near supplying an answer to these problems, and their resolution will be difficult.

All known *Haplochromis* species and the monotypic genera show clearly defined sexually dimorphic coloration. With two exceptions, it is the male fish that has the brighter and more colourful livery, even in non-breeding periods. The exceptions are provided by *H. dichrourus* (Greenwood, 1967) and *H. chromogynos* (Greenwood, 1959b). In *H. dichrourus*, although the sexes are differently coloured, the female is more polychromatic than the male (see Greenwood, 1967 ; these observations have since been confirmed and extended by additional specimens). In *H. chromogynos*, too, there is dimorphic coloration. Males have what may be termed 'typical male coloration' but the females are distinctly marked by a black-and-white piebald coloration that is otherwise found as a sex-limited polymorph in females of certain species (where its frequency is never in excess of *ca* 30 per cent [Greenwood, 1956a and b, and p. 53 below ; also discussion in Fryer & Iles, 1972]).

The importance of male coloration in species recognition amongst many cichlid genera is now well established (see Baerends & Baerends van Roon, 1950). Thus it is not surprising to find male breeding coloration is, at least to the human eye, species specific. (Personal observations ; see also descriptions in Greenwood, 1956–69.)

With the relatively limited chromatophore pigments and interference colours available it is also not surprising to find these differences not always sharply marked. That is, there is not always a gross interspecific difference in coloration or colour pattern. Instead, apart from differences in ground coloration, the specific differentiae involve differently coloured suffusions (red, orange, coppery) over various parts of the flanks, belly, chest or head, and differently patterned colour flushes (especially red or pink) on the median fins (particularly the anal and caudal fins). In addition there may be coloured lappets (predominantly red or orange) on the dorsal fin, or these may be colourless (Plate 1).

It is difficult to create a word picture for the ground colour of the body. Reduced to basic colours, four principal types are found, *viz* : blue-grey (the commonest), shades of green and blue (from turquoise to malachite), yellow to yellow brown (with beige as an extreme) and, least common, uniformly black.

Ground coloration is most intense on the dorsal half of the body, shading to silver, grey or yellow ventrally ; in other words the fishes are basically countershaded. Most species show patterns of dark vertical bars and horizontal stripes on the body, and various bars and stripes on the head and opercular region (see Text-fig. 8). These patterns are, with few exceptions, under emotional control and seem to play an important part in the fishes' repertoire of signals (see Wickler, 1964). The exceptional species (e.g. *H. percoides* [Text-fig. 52A], *H. flavipinnis* [Text-fig. 52B], *H. martini* [Text-fig. 50] and *H. squamulatus* [Text-fig. 49] and some others) have fixed, but individually variable, patterns which are present in both sexes.

Species from deepwater habitats also show bright male coloration (Greenwood & Gee, 1969). Indeed, several species have colours not seen among inshore species,

PLATE 1

Above. *Haplochromis riponianus*. Sexually active male. Note the prominent 'egg dummies' on the anal fin.

Below. *Haplochromis brownae*. Adult male, showing almost complete development of breeding coloration.

Photographs by André Roth.

PLATE I

for example bright purple and pastel green. Although various shades of purple seem to predominate in the species so far described, this colour does not appear to be at all common in the species still awaiting description.

Females from all habitats, in sharp contrast to males, are drably coloured, with greyish silver or sandy ground colours predominating (*H. chromogynos* and *H. dichrourus*, excepted ; see p. 52 above). Unlike males there are few marked interspecific differences in female coloration, but slight differences do exist between some species. Cephalic and somatic bars and stripes are developed as in males, and also seem to be under emotional control.

A very prominent and characteristic feature in the males of all *Haplochromis* species (and related genera) are the ocelli or 'egg dummies' on the anal fin. From one to several 'egg dummies' are located on the posterior quarter of this fin. Each is a roughly ovoid colour patch, usually yellow or orange-yellow, surrounded by a completely colourless ring of fin membrane. In life this clear zone gives the spot a three dimensionality such that it closely resembles a *Haplochromis* egg. The functional importance (and indeed the name) of these 'egg dummies' was discovered by Wickler (1962a and b, 1968). He was able to demonstrate that, after the female has oviposited and taken the ova into her mouth, the male displays his anal fin in such a way as to bring these spots into prominence. The female then attempts to pick up the spots, presumably identifying them as newly deposited ova. Whilst she is in the close proximity of the male's anal fin, and thus near his genital papilla, he ejaculates (Wickler, *op. cit.*). Clouds of sperm are then drawn into the female's mouth, and the ova are fertilized. In other words, these fishes have evolved a kind of secondary internal fertilization.

'Egg dummies' have not been found in any female, but there are often colour spots on the anal fin of females in the same position as the 'egg dummies' in males. However, there is never a clear zone around the spots and they certainly do not have the appearance of a true 'egg dummy', neither are they so large nor so clearly defined.

Sex-limited female polychromatism occurs in at least eight species of the Lake Victoria *Haplochromis*, and in two of the endemic monotypic genera (*Macropleurodus bicolor* and *Hoplotilapia retrodens*). In none of these species does the occurrence of polymorph individuals exceed *ca* 30 per cent of females in a population. The commoner polymorph coloration found in all species showing the phenomenon is an irregular black piebald on a silver or yellow background (Text-fig. 29) ; no intergrades between the piebald and normal female coloration have been found. This piebald pattern does not vary with the emotional state of the fish, nor with the type of habitat in which the fish is living. A second morph is also known from *Hoplotilapia retrodens*, and in at least one undescribed *Haplochromis* species ; it comprises a blotched yellow and orange ground coloration on which is superimposed small black blotches or merely a fine but distinct peppering of melanophores (Greenwood, 1956a).

The eight *Haplochromis* species showing piebald coloration belong to at least four different phyletic groups, none of which shows especially close interrelationships within this admittedly closely related species flock. Algal grazing species (*H. nigricans*), paedophages (*H. cronus*, *H. obesus* and *H. barbarae*), mollusc eaters (*H.

sauvagei), specialized insectivores (*H. chilotes*) and a piscivore (*H. altigenis*) all exhibit the phenomenon.

Selective advantages associated with the maintenance of this balanced polymorphism are unknown (Greenwood, 1956a ; see also discussion in Fryer & Iles, 1972), but must be fairly substantial considering both the frequency of polymorph individuals in a population, and the fact that it occurs in so many species.

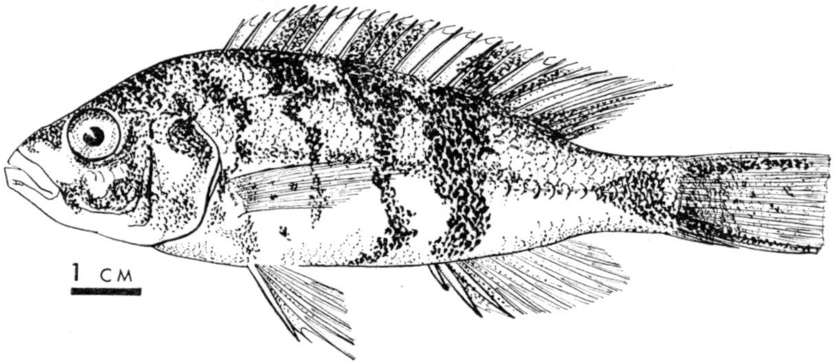

FIG. 29. *Haplochromis barbarae* female, showing piebald coloration (see p. 53).

The usually complete linkage of piebald coloration to the female sex is disturbed in *Macropleurodus bicolor*, where two piebald males have been found (Greenwood, 1956a). In one of these fishes the pattern is identical with that of females, but in the other it is less intense and the background coloration is darker. As far as I can determine, the gonads of these atypical males contained only testicular tissue, and provided no evidence to suggest possible protandry in the species (as is suspected for a Malawian genus *Labeotropheus fuelleborni* [report by Professor H. Peters, quoted in Fryer & Iles, 1972 : 172]). A male showing partial female-type piebald colours is also recorded for *Hoplotilapia retrodens* (Greenwood, 1956a), and I have seen a live *Haplochromis chilotes* with piebald coloration and well-defined 'egg dummies' on the anal fin. Unfortunately it was not possible to dissect the latter specimen, an aquarium fish.

The genetic basis for this almost completely sex-limited polychromatism, and its occasional breakdown, has not been determined. Fryer & Iles (1972) argue cogently in favour of its being the manifestation of a potency balance between the expression of alleles on autosomes and 'suppressor' alleles on the sex chromosomes. On the assumption that female cichlids are the heterogametic sex (which is also assumed in my explanation of the phenomenon ; Greenwood, 1956a), this interpretation seems at least to provide a working hypothesis. I am less sanguine about their view that the occurrence of atypically coloured (i.e. piebald) males violates the sanctity of male coloration as an important factor in species recognition. That the proportion of piebald males is very much lower than that of polymorph females suggests that alleles with a potency sufficient to overcome the 'suppressor' alleles are extremely rare.

Could not this rarity be attributable to adverse selection pressure on those individuals possessing them? That is, the chances of such males being 'recognized' and 'chosen' by a female (an obligatory prelude to spawning) are greatly reduced; this in turn would lead to a great reduction in the number of these 'powerful' alleles in the population. Is it not even possible that piebald males do not mate at all, and that the alleles in question are maintained purely by mutation?

To me, a far more telling point against the importance of male coloration as the prime means of species recognition in courtship derives from the *Haplochromis* species flock in Lake George. In this lake the euphotic zone extends only to a depth of 50–60 cm (Greenwood, 1973a). Most *Haplochromis* species live and apparently spawn below this depth, or at least in places with extremely low light intensities. To the human observer underwater in these lighter areas, a fish's shape, but not its colour, is just apparent. Thus in these species there seems to be a *prima facie* case against the overriding importance of male coloration. Yet, males are brightly coloured, and their coloration (with one exception, see Greenwood, 1973a) is species specific, and does not show any greater degree of intraspecific variability than do the colours of the Lake Victoria species.

Finally, in dealing with colour aberrancies in Lake Victoria species, mention must be made of apparently melanic *Haplochromis obesus*. Atypically dark individuals of both sexes are known, but melanic males are commoner. The degree of darkening shown by these fishes is somewhat variable; extreme individuals are uniformly black, but others are much brighter ventrally and do in fact intergrade with the normal male coloration., No correlation has been observed between locality or environment and the presence of melanic individuals, nor with the degree of melanism displayed.

In a large and complex species flock like that of Lake Victoria, with many species occupying the same habitats, one might well expect a number of interspecific hybrids. Detecting hybrid individuals on purely morphological grounds would be difficult because of the close similarity between putative parent species. Nevertheless, despite constant awareness of this possibility, I have found only one specimen that might be a hybrid. Whatever the barriers to interspecific mating may be, they seem to be very effective.

Comparisons between the coloration of *Haplochromis* species from Lakes Victoria and Malawi are difficult to make, partly because of difficulties in verbalizing colours, and partly because colour descriptions are not available for many Lake Malawi species. Nevertheless, the views of those who know both faunas strongly indicate a fundamental difference in colour patterns and dominant colours (see remarks in Fryer & Iles [1972]; also personal comments by Drs Fryer, Lowe-McConnell and Trewavas).

The reasons for these differences are probably manifold (for example, the greater number of pelagic species in Lake Malawi, the existence of shoaling species in that lake etc.), and must be, in part, of phylogenetic origin. In this context it is important to emphasize the basic similarity in coloration between species of the Lake Victoria and Lake Edward–George fishes, where phyletic integrity is ascertainable on other grounds (Greenwood, 1973a).

INTERRELATIONSHIPS OF THE LAKE VICTORIA *HAPLOCHROMIS* SPECIES

Having outlined the main morphological and biological features of this *Haplochromis* species flock, attention can now be turned to four major evolutionary questions. These are, the origin of the flock, its internal phylogeny, the factors underlying its explosive speciation and adaptive radiation and, finally, why it is that, under such conditions the Cichlidae, more than any other family, react in this fashion (see p. 4 above and pp. 103–111 below).

The phylogenetic problem can be considered first (Text-fig. 70). Any attempt to analyse this flock is fraught with difficulties. The very basis for a phyletic study, the flock's monophyletic origin, cannot be established unequivocally. Reasons for assuming a mono- or oligophyletic origin for the endemic species have already been discussed (p. 20 above). To these may be added one other, namely, that a careful survey of most known species indicates a basic morphological homogeneity that would be unlikely if several unrelated species were implicated as ancestors. In other words, if the flock is not strictly monophyletic in origin, its stem species were likely to have been no more distantly related than sister species (*sensu* Hennig, 1966). Be that as it may, it is certainly possible to establish monophyletic groups within the flock (Text-fig. 70).

At a gradal level, the Victoria *Haplochromis* species present an interesting and, for extant animals, an unusual picture of virtually complete morphological intergradation between the generalized and the specialized in any one adaptive radiation. Two trenchant examples are found in the mollusc-eating species and a third in the piscivorous predators (see pp. 31 and 37 above).

Graded anatomical stages, involving changes in dentition and dental pattern among species that shell molluscs orally are all represented in the lineage *H. sauvagei*, *H. prodromus* and *H. xenognathus*, with *H. granti* a slight deviant from the *H. prodromus* stem (see Text-fig. 41 ; and Greenwood, 1957). The genus *Macropleurodus* is an extreme modification of the same morphotype (see Text-figs 75A and B ; also Greenwood, 1957). Stages in the hyperdevelopment of a pharyngeal crushing mill amongst mollusc eaters are well represented by seven species, *H. pallidus*, *H. riponianus*, *H. saxicola*, *H. humilior*, *H. theliodon*, *H. obtusidens* and *H. ishmaeli*, the species listed in approximate order of increasingly molarized lower pharyngeal teeth (see Text-fig. 5 and Greenwood, 1960 ; also p. 37 above). The piscivorous predators present an even more complex and just as complete picture of bridged gaps between specializations (see Greenwood, 1962, 1967, 1973b). The other trophic groups all show similar, but often shorter, trends in specialization.

The question now to be asked is whether all the constituent species of each gradal complex represent truly monophyletic lineages within the flock, or whether we are confronted by a web of parallelisms. Detailed studies of the species comprising the different radiations suggest that the latter explanation is the more likely one. The task then is to untangle the gradal groups by seeking out combinations of specialized characters (apomorph characters of Hennig, 1966) that will link together phyletic lineages within each grade.

The student of the Lake Victoria flock is in the position of a palaeontologist who has a nearly complete phyletic record for several branches of a monophyletic lineage, the extant end-products of which are clearly defined taxa, but some of whose ancestors tried out, as it were, specializations now characterizing living taxa in a sister lineage. Unlike the palaeontologist, the student of the Lake Victoria *Haplochromis* has no temporal sequence to help him sort out branching points.

Historically, some of these difficulties could spring from the flock's relative youth. In an older assemblage a number of species now seen as representing an intermediate stage of specialization could well have been eliminated as unsuccessful competitors with more specialized species of their own or related lineages.

Whether or not an older cichlid species flock would be an easier subject for phyletic analysis, and whether the results would be more precise, are moot points. Unfortunately no such test has been made on the flocks of either Lake Tanganyika or Lake Malawi in their entirety. Fryer (1959; also repeated in Fryer & Iles, 1972) has, however, attempted to work out the interrelationships of the 'Mbuna' generic complex in Lake Malawi. His criteria of relationship are not explicitly defined and the resulting 'tree' seems more an indicator of parallel radiation than of phylogenesis.

A feature shared by *Haplochromis* in both the older flock of Lake Malawi and the younger one of Lake Victoria is the existence of two or more very similar species clustered around any one level of adaptation. It is these species in particular that have given the Cichlidae a reputation among taxonomists of being a difficult group. Examples of species knots from Lake Victoria would include five species in the insectivore lineage (*H. macrops*, *H. megalops*, *H. pallidus*, *H. lacrimosus* and *H. piceatus*; see Text-figs 3, 8 and 9 and p. 30 above), the five species of the crustacean eating, deepwater, *H. tridens* group (see p. 39 above; also Greenwood & Gee, 1969), *H. sauvagei* and *H. prodromus* in the mollusc eating grade (Text-figs 3 and 19–21; see also p. 37 above; and Greenwood, 1957), *H. ishmaeli* and *H. pharyngomylus* in the other molluscivore grade (see p. 37 above; and Greenwood, 1960), *H. riponianus*, and *H. saxicola* (Text-figs 3 and 42) also mollusc crushers (Greenwood, *op. cit.*), *H. obesus*, *H. maxillaris* and *H. melanopterus* among the paedophages (Text-figs 3 and 17; see also p. 33 above; and Greenwood, 1959b), and many other examples of twin, triplet and even quintuplet species among the piscivorous predators (Greenwood, 1962, 1967).

These species groups (if they are truly phyletic, and some certainly seem to be that) may reflect the physiographical background to speciation in the lake, and be the products of simultaneous multiple speciation each from the same common ancestor isolated in different water bodies (see p. 10 above and p. 114 below).

Relatively few characters can be used to construct phylogenies. It will be recalled that meristic characters are almost uniform throughout all species. Thus, interest centres on cranial and dental features which, on the whole, clearly show levels of specialization or generalization.

As the base line for this study I have identified as generalized those characters shared by Victoria species and a fluviatile species widespread in the rivers of Uganda (i.e. a taxon anatomically identical with, if not actually *H. bloyeti*; see Greenwood, 1971).

Skull shape, jaw form and tooth morphology in this species are illustrated in Text-figs 30 and 31. Characteristic features of this skull type are the relatively decurved dorsal profile to the preorbital face of the neurocranium (giving the orbital margin a near-rounded circumference), the high cranial vault and the relatively short ethmovomerine region. The premaxillary ascending process is shorter than the dentigerous arm of the bone, and the lower jaw (dentary plus articular) is neither foreshortened nor elongate.

The lower pharyngeal bone is not noticeably thickened, and its equilateral dentigerous surface is covered by fairly well-spaced rows of cuspidate and laterally compressed teeth. Teeth in the median rows are usually a little coarser than those situated laterally (Text-fig. 30A).

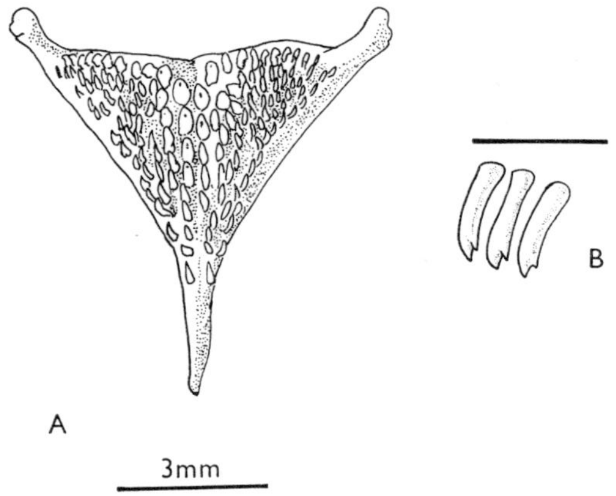

FIG. 30. *Haplochromis bloyeti*. A: Lower pharyngeal bone in occlusal view. B: Outer row teeth from the premaxilla; viewed anterolaterally. Scale = 1mm.)

The outer jaw teeth in *H. bloyeti* are unequally bicuspid, the cusps triangular in outline (Text-fig. 30B); the inner teeth are small and tricuspid, and are arranged in not more than three rows in either jaw. Usually a few unicuspid teeth occur posteriorly in the outer row of the premaxilla, and in larger fishes (> 80 mm long) a few unicuspids are intercalated with the bicuspids anteriorly and anterolaterally in both jaws.

Among extant Lake Victoria endemic species, *H. pallidus* has a syncranium and dentition virtually identical with that of *H. bloyeti* (see Text-figs 31–33), and there are several other similarly generalized species known.

Having established the anatomical nature of a generalized species, each trophic group (see pp. 58–80) will be considered in turn, probable phyletic lineages (based on shared specialized characters, i.e. synapomorphy) will be delimited and, where possible, interrelated.

The insectivorous species

Typically generalized skull, jaws and dentition are found in five species (*H. pallidus* [Text-figs 32 and 33], *H. macrops* [Text-figs 32 and 33], *H. lacrimosus*, *H. megalops* and *H. piceatus*; see Greenwood, 1960 and Greenwood & Gee, 1969). In all except *H. pallidus* the lower pharyngeal dentition is also generalized; in *H.*

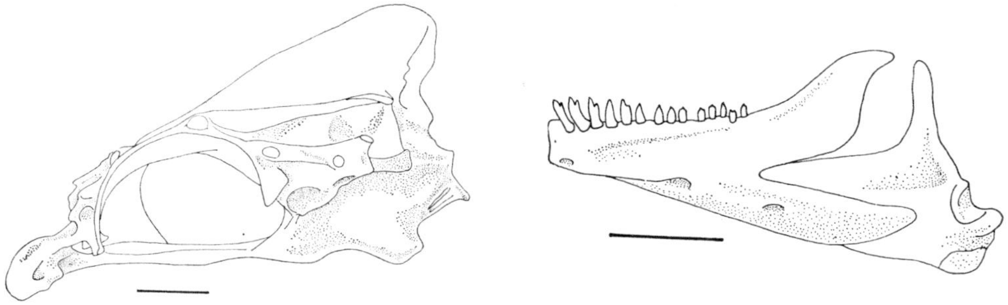

FIG. 31. *Haplochromis bloyeti*. Neurocranium and lower jaw, in left lateral view. (Scale = 3 mm.)

pallidus some median teeth are slightly enlarged. A sixth species, *H. cinereus*, is osteologically one of this group, but differs in having a predominance of unicuspid over bicuspid outer jaw teeth, even in fishes at a size where bicuspids predominate in other species (Greenwood, 1960).

Because of their generalized cranial anatomy it is impossible to determine the interrelationships of these six species ; all could be cognate. However, except for *H. pallidus* all species have the preorbital skull face slightly more decurved than in the presumed ancestral type, with which *H. pallidus* is virtually identical.

Anatomically, it is interesting to see amongst these fishes a representation, in embryo as it were, of many characters that are developed in various characteristic ways among more specialized trophic groups. For instance, there are the unicuspid teeth of *H. cinereus* (but see below), the incipient development of enlarged pharyngeal teeth in *H. pallidus*, and a basic syncranial 'bauplan' that, through differential growth of certain elements (or regions in the case of the neurocranium), provides the starting point for the different types found in other trophic groups.

None of the species considered so far has adults exceeding a length of more than 105 mm.

Haplochromis saxicola is, superficially, like the foregoing species (*cf* Text-figs 3 with 8 and 9). However, its neurocranial morphology departs from the basic type towards that of some piscivorous species (Text-figs 34 and 68). That is, the preorbital region is more elongate (as is the entire skull anterior to the brain case), its dorsal profile is not noticeably decurved and it slopes downwards less steeply, and there is a reduction in the depth of the brain-case region. The jaw dentition too differs somewhat from the generalized type because unicuspid teeth predominate ; since the smallest *H. saxicola* known is larger than the largest members of the generalized group, this observation may be of doubtful significance (see p. 106). Thus, *H. saxicola*, although still an insectivore, does depart from the presumed basal members of that trophic group in several important anatomical features, and in reaching a larger adult size, *viz* 125 mm.

The nearest living relative of *H. saxicola* is *H. riponianus* (Text-figs 34 and 42), here classified as a mollusc eater (see below). Nevertheless, as the representative of a specialized structural level, the *H. saxicola* condition could be ancestral to the piscivorous predator radiation (see p. 82).

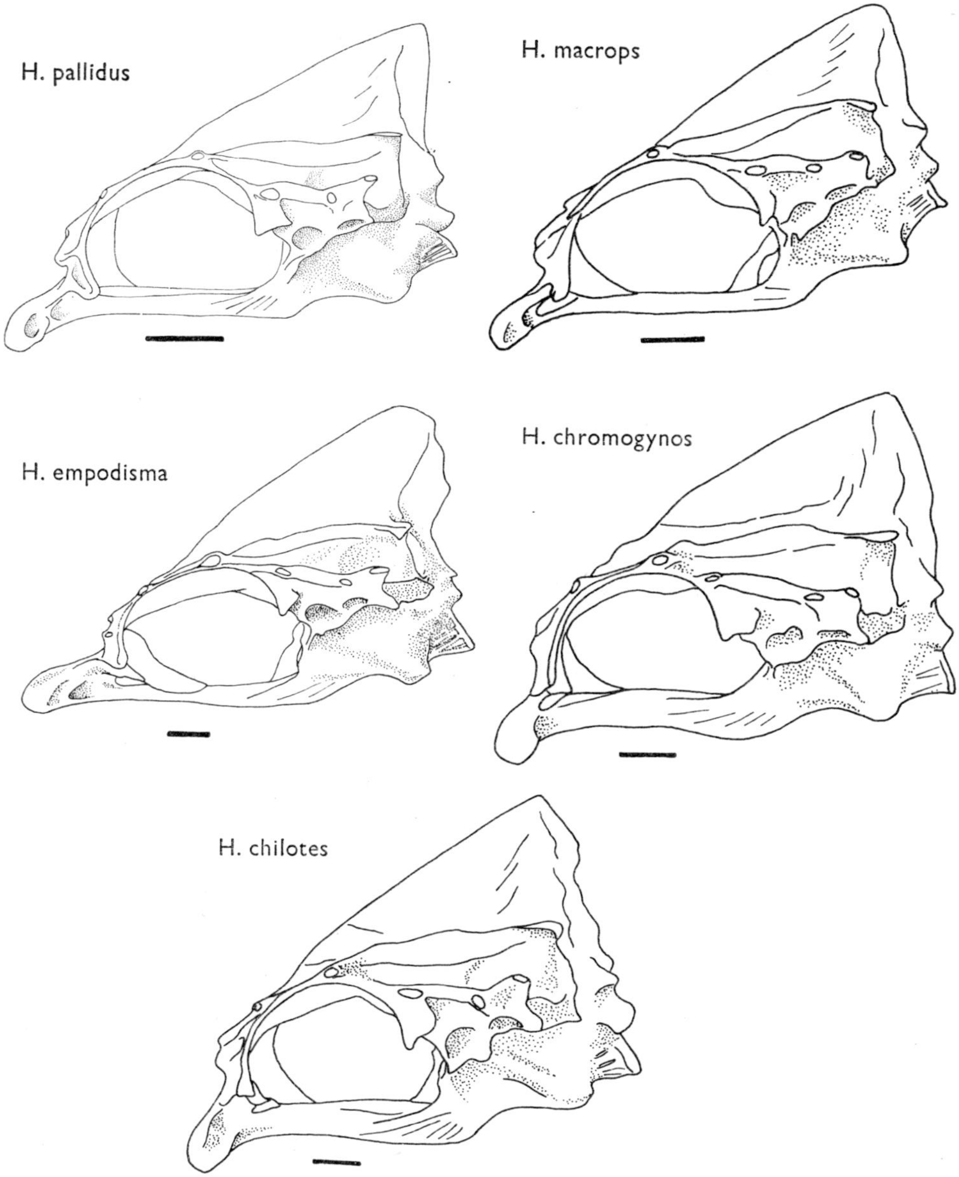

FIG. 32. Neurocranial form in insectivorous *Haplochromis* species (see also Text-fig. 33). (Scale = 3 mm.)

Also related to *H. saxicola*, at least on the basis of jaw and neurocranial morphology (Text-fig. 34), is *H. aelocephalus*, a predominantly insectivorous species that also feeds on other invertebrates, including molluscs. Externally, *H. aelocephalus* is readily distinguished by its protracted snout which, in extreme individuals, is almost tubular (see Greenwood, 1959b). The outer jaw teeth in *H. aelocephalus* are simple unicuspids like those in *H. saxicola*, but the inner teeth are arranged in broad bands, more like those in the *H. sauvagei–H. chilotes* lineage (see p. 72 below). There is, however, no reason to suppose that *H. aelocephalus* is at all closely related to that lineage, from which it differs in neurocranial morphology, the shape of the outer jaw teeth, and in having an elongate and slender lower jaw.

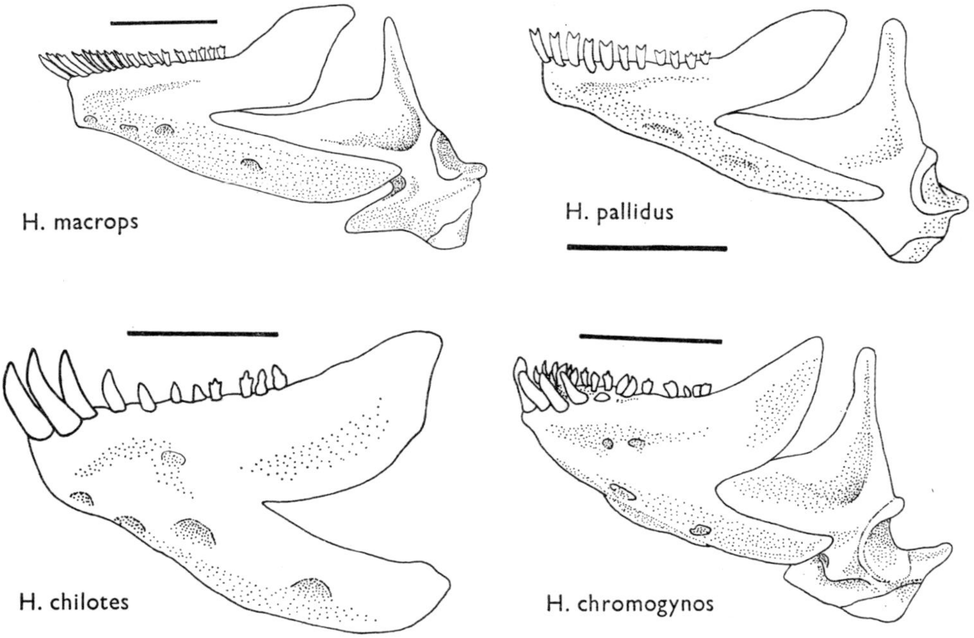

FIG. 33. Lower jaw form in insectivorous *Haplochromis* species (see also Text-fig. 32). (Scale = 3 mm.)

Even superficially, *H. chilotes*, with its hypertrophied and lobate lips (Text-fig. 10), would qualify as a specialized species, a rating confirmed by its feeding habits (see p. 31). The oral dentition of *H. chilotes* departs markedly from that in all the species so far considered (Text-fig. 33). The outer teeth are stout unicuspids with the crown strongly incurved; such teeth occur in only one other group of species (see p. 72 below). The lower jaw is short and stout, but the ascending process of the premaxilla is relatively longer than in the generalized type of bone. The neurocranium of *H. chilotes* could be classified as a derivative of the generalized type in which the preorbital region has become strongly decurved. Adult *H. chilotes* reach a length of *ca* 150 mm.

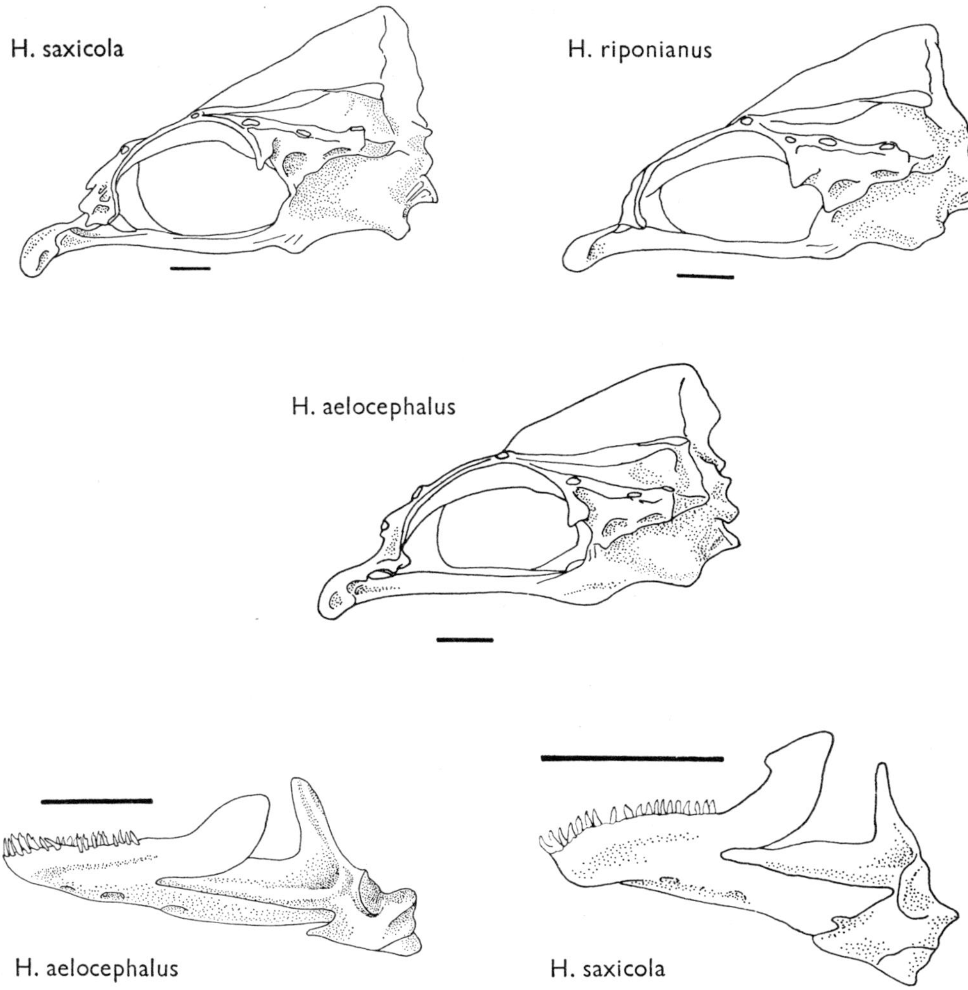

FIG. 34. Neurocranial and lower jaw form in insectivorous *Haplochromis* species. (Scale = 3 mm.)

Virtually identical syncranial morphology and dentition are found in *H. chromogynos* (Text-figs 32 and 33), but the lips are not hypertrophied in this species ; adult fishes are apparently smaller (maximum length 110 mm) than in *H. chilotes*.

I would consider *H. chilotes* and *H. chromogynos* to be sister species, with *H. chilotes* the derived (apomorph) member of the pair. Similarly specialized skull and jaw forms and outer tooth morphology are seen in one lineage of mollusc eaters, and it is thought that *H. chilotes* and *H. chromogynos* are members of the same lineage (see p. 72, and Text-fig. 66).

Finally in the insectivorous group there is *Haplochromis empodisma*, a species which could be described as a deep-bodied and larger form of the five species described first in this section (Greenwood, 1960) ; adults of *H. empodisma* reach a length of 120 mm, compared to 100 mm in the other species.

The oral and pharyngeal dentition of *H. emposdisma* are of a basic type, but the neurocranium departs quite noticeably both from the generalized type (Text-fig. 32) and, especially, from that of *H. chilotes* and *H. chromogynos*. The preorbital profile is straighter and slopes less steeply because the brain-case depth is somewhat shallower, giving the whole skull a more linear and less rounded appearance. In fact, the skull of *H. empodisma* is almost intermediate between the *H. saxicola* type and that of the more generalized species (e.g. *H. pallidus* and *H. macrops*).

Admittedly the *H. empodisma* neurocranial form differs less markedly from the basic form (e.g. *H. pallidus*) than does the skull in *H. saxicola* or, more especially, *H. chilotes*. But it does seem to represent a distinct type and one which appears elsewhere within the flock. One near relative of *H. empodisma* is the mixed mollusc—insect eating *H. obtusidens* (see p. 73 below ; and Greenwood, 1960).

To summarize. Among the insectivorous species there are four distinct lineages (Text-figs 65, 66, 68 and 70). One, the most generalized, departs little from the presumed ancestral and fluviatile species represented by *H. bloyeti*. The second (*H. saxicola*) shows an elongation of the skull and jaws approaching that found in certain piscivorous and paedophagous species. The third lineage (*H. chilotes* and *H. chromogynos*) exhibits strong decurvature of the skull anteriorly, strengthening of the jaws (especially the lower) and a distinctive dentition, all features shared with a specialized branch of the mollusc-eating trophic group. The fourth lineage (*H. empodisma*) represents a slight departure from the generalized *H. pallidus*-like condition towards a type seen in certain piscivorous predators and other trophic groups.

Apart from their dentition, the differentiation of these lineages is manifest in a changed shape to the anterior moiety of the neurocranium and, to a lesser extent, in the relative length of the lower jaw. In other words, a change in relative growth patterns. It is probably significant in this connection that there are marked differences in the maximum adult size attained by members of the various lineages, with those of the more specialized lines growing to a larger size.

Phytophagous species

In most respects the morphologically least differentiated member of this trophic group is *H. phytophagus* (see Greenwood, 1966b), one of the two species known to feed directly on macrophytes ; see pp. 39 & 41. Syncranial organization in *H. phytophagus* is like that in any generalized insectivore. Specialization is seen in the coarser jaw teeth, the coarser, less numerous teeth on the pharyngeal bones, and in the lengthened intestine. The dental modifications, however, are but slight variants of the basic bicuspid oral dentition, and even slighter changes in the pharyngeal dentition ; both are adaptations for biting and then macerating the plant tissues eaten (particularly leaves).

Three of the four algal grazing species (*H. nuchisquamulatus*, *H. lividus* and *H. obliquidens*) also have a generalized syncranium. All four species (i.e. including *H. nigricans*, see p. 39) have a greatly elongate intestine (Greenwood, 1956b) and rather

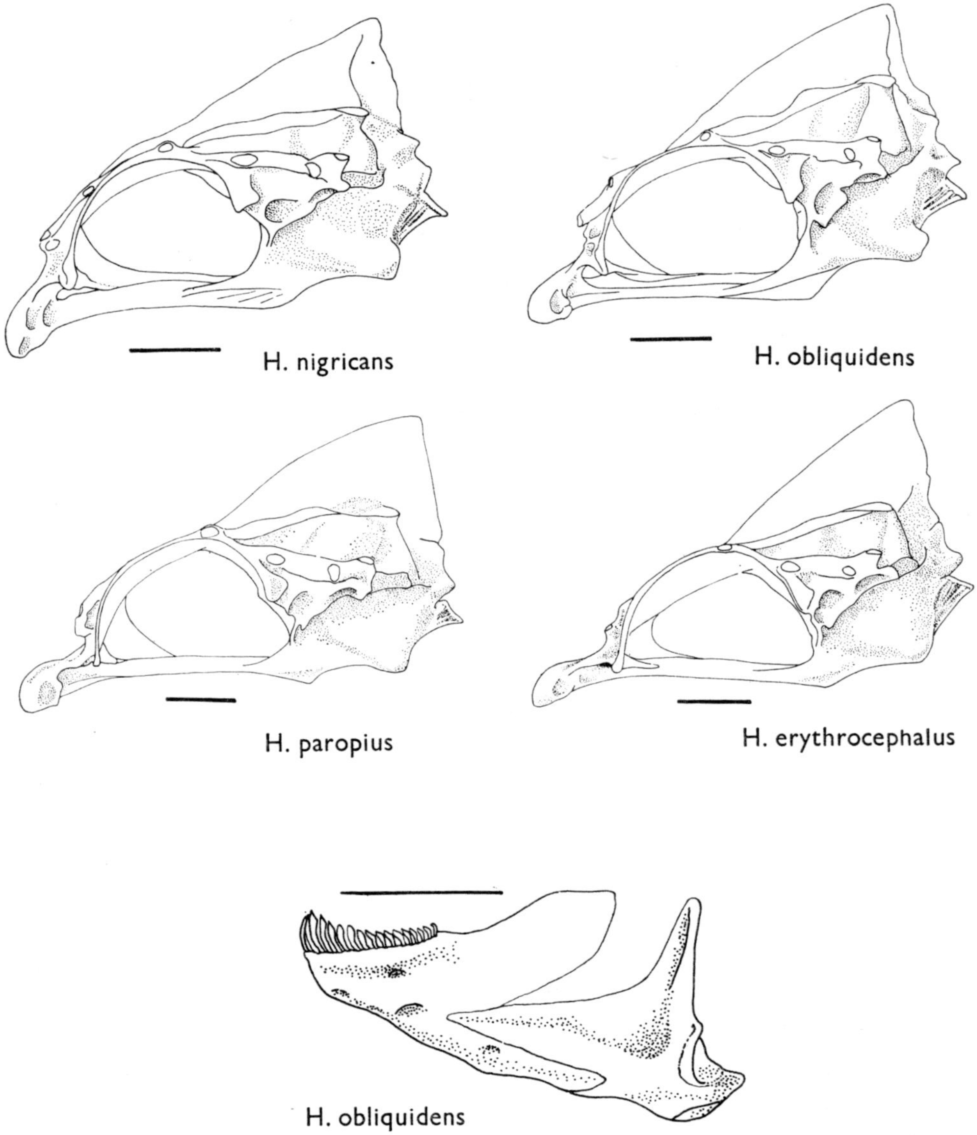

FIG. 35. Neurocranial and lower jaw form in phytophagous *Haplochromis* species. (Scale = 3 mm.)

fine and numerous pharyngeal teeth, but *H. nigricans* does show certain syncranial pecularities not shared by the others.

As far as tooth shape is concerned, *H. nuchisquamulatus* has the least specialized dentition (Text-fig. 4E) since it is of the basic bicuspid type with acutely pointed cusps ; compared with the generalized insectivore, however, *H. nuchisquamulatus* does show an increase in the number of inner tooth rows, a wider area of implantation for these teeth, and all the teeth are moveably implanted. *Haplochromis lividus*, in comparison, has many outer jaw teeth in which the major cusp is relatively broader and obliquely truncate (not acute), see Text-fig. 4F ; a few teeth retain a generalized shape. Inner jaw teeth and the pharyngeal dentition in *H. lividus* do not differ from those of *H. nuchisquamulatus*.

The morphological trend apparent in the teeth of *H. lividus* is accentuated in *H. obliquidens*. Here the minor cusp is suppressed, and the major cusp is drawn out and obliquely truncate (Text-fig. 4G) ; occasionally some teeth have a vestigial minor cusp, and a few teeth of the *H. lividus*-type may be present posteriorly in the premaxilla. Generally the inner teeth are typical tricuspids, but there is often an admixture of these and teeth differing from those of the outer row only in their smaller size. (It may be noted that larval *H. obliquidens* have the fine, setiform teeth that probably occur in all *Haplochromis* species irrespective of their definitive adult dental morphology ; Greenwood, 1956b, and further, unpublished observations.)

It seems reasonable to consider *H. obliquidens*, *H. lividus* and *H. nuchisquamulatus* as members of a phyletic lineage ; certainly *H. obliquidens* and *H. lividus* are sister species. Although it is difficult to be certain of the relationship between these species and *H. nuchisquamulatus*, a relationship seems to be indicated by the slight departure of tooth form in the latter towards the *H. lividus* type (i.e. moveable implantation, expansion of the crown and its less acutely pointed tip). The base of this lineage lies undetectably within the generalized insectivore species group.

The relationships of *H. nigricans*, the fourth species feeding on phytoplankton, provide something of a puzzle. Its oral dentition is of a modified bicuspid type (Text-fig. 36), but does not show the obliquely cuspidate major cusp typifying the

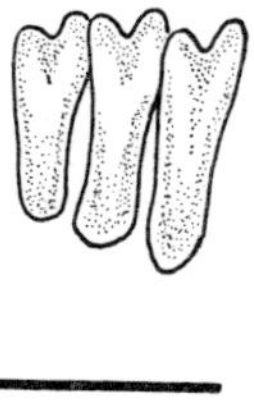

FIG. 36. Outer row teeth from the dentary of *H. nigricans* (seen in lateral view). (Scale = 1 mm.)

other species. Indeed, many teeth are subequally cuspidate, the cusps rather bluntly pointed, although the whole crown, as in the grazers, is expanded relative to the neck of the tooth (Text-fig. 36). The inner teeth are tricuspids and are arranged in numerous rows with a reduced space between them and the outer tooth row (again, like the other species). *Haplochromis nigricans* differs in the shape of its neurocranium, in which the preorbital face is somewhat decurved (Text-fig. 35), and in having a stouter lower jaw. In both these characteristics, *H. nigricans* shows some approach to the *H. chilotes–H. chromogynos* condition.

The question raised by these contrasting features is whether the neurocranial and jaw characters indicate relationship to the *H. chilotes* lineages, or whether they are to be interpreted as parallelisms.

I incline towards the latter interpretation mainly because of the very different morphology of the teeth in *H. chilotes* and other members of its lineage. *Haplochromis nigricans*, unlike the other periphyton grazers, feeds principally from rocks and not plants, and it has a rather different feeding method. The epiphytic feeders scrape algae mainly from leaves which are taken and held between the jaws. *Haplochromis nigricans*, on the other hand, nibbles algae from a rigid substrate, for which the stout jaws, downward protrusion of the upper jaw (because of the decurved surface on which the premaxillae slide) and the subequally bicuspid outer teeth would appear to be adaptations.

The three remaining phytophages (*H. cinctus* [Text-fig. 27], *H. paropius* and *H. erythrocephalus* [Text-fig. 26]; Greenwood & Gee, 1969) all feed on phytoplankton apparently gathered from the bottom and not while in suspension; all have relatively long intestines.

Unfortunately, lack of material precludes a detailed knowledge of cranial anatomy in *H. cinctus*; however, its dentition, both oral and pharyngeal, is of the generalized bicuspid type (Greenwood & Gee, 1969). *Haplochromis paropius* has the syncranial architecture and oral dentition of a generalized insectivore like *H. macrops* (see Text-figs 32, 33 and 35), but its pharyngeal dentition is of the finer type found in the periphyton feeders (Greenwood & Gee, *op. cit.*).

The neurocranium in *H. erythrocephalus* differs from this generalized type in exactly the same way as does the neurocranium of *H. empodisma* (see p. 63 above; also Greenwood & Gee, *op. cit.*). The pharyngeal and oral dentition of these two species is also similar. No immediately obvious adaptive features are apparent in this type of syncranial architecture. Possibly the larger mouth and deeper oropharyngeal cavity (relative to those of the generalized type) are of advantage to a species which, when feeding, must pass a considerable volume of water and particulate material through this cavity in order to obtain its food.

Because of their generalized cranial characters and dentition, *H. cinctus* and *H. paropius* cannot be related to any particular species. But because of their specialized alimentary characters the species must be considered as apomorph derivatives of the generalized insectivore stock, and are probably related to one another (i.e. sister species). *Haplochromis erythrocephalus*, on the other hand, can be related to *H. empodisma* also as a derived species (Text-figs 65 and 70).

To summarize: Three phyletic lines are represented among the phytophages so far considered. The periphyton grazers comprise a single lineage in which most of the adaptational stages of the radiation are still extant, together with a differently specialized offshoot, *H. nigricans*. The phytoplankton feeders comprise two lineages; one (*H. cinctus* and *H. paropius*) is little different from the basal insectivore stem, the other (*H. erythrocephalus*) is part of a lineage whose living basal representative is probably *H. empodisma*. *Haplochromis phytophagus,* a browser on macrophytes, could belong to the same lineage as *H. cinctus* ; like those species its dentition shows little departure from the generalized type, but it does have the same alimentary canal specializations. Possibly the three species are an offshoot of the lineage culminating in the periphyton grazers (Text-figs 65 and 70).

There remains one other species, *H. acidens* (Greenwood, 1967), which, like *H. phytophagus,* feeds directly on macrophytes. The skull, jaws and unicuspid dentition (Text-fig. 37) of this species conform with the type found among a group of specialized piscivorous predators (Greenwood, *op. cit.*), as does the pharyngeal dentition. Furthermore, despite the vegetarian habitats of *H. acidens,* its intestine is relatively shorter than in other plant-eating species, although it is longer than in a piscivore.

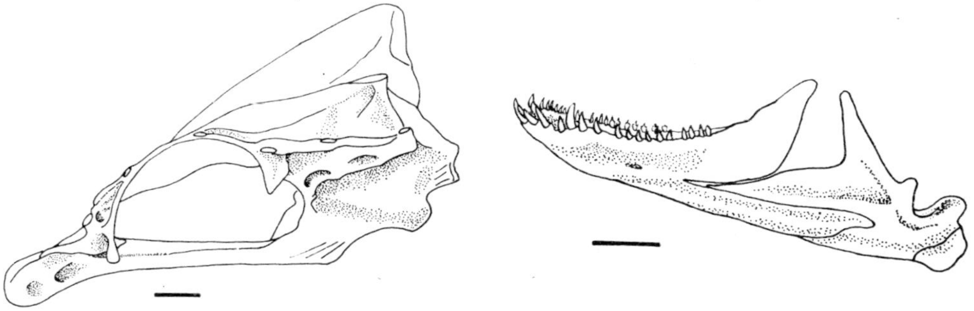

FIG. 37. Neurocranium and lower jaw of the enigmatic *H. acidens* (see p. 41). (Scale = 3 mm.)

I have suggested elsewhere (Greenwood, 1967) that, anatomically, *H. acidens* could be a member of a particular piscivore lineage (the '*serranus*' group, see Greenwood, 1962). It could also be a derivative from an *H. empodisma*-like species ; the principal change involved would be an increase in the relative number of unicuspid teeth so that these teeth predominate over bicuspids in both jaws (there is an admixture of bi- and unicuspids in large *H. empodisma* [i.e. > 95 mm long] and in *H. acidens* < 90 mm in length). In both species the teeth, irrespective of cusp form, are slender in comparison with those of the generalized type. Considering the rather indefinite feeding habits of *H. empodisma* (a benthic insectivore also ingesting quantities of plant debris) and the interspecific similarities in syncranial architecture, I would now consider *H. acidens* to be a derivative of the *H. empodisma* lineage, rather than of the '*serranus*' piscivore group (see also p. 41).

Species feeding on benthic Crustacea

This small and trophically rather ill-defined assemblage of five species, the '*tridens*' group (*H. dolichorynchus, H. tyrianthinus, H. chlorochrous, H. cryptogramma* and *H. tridens* [Text-figs 24 and 25 ; Greenwood & Gee, 1969 ; Greenwood, 1967] is, however, well-defined morphologically. Group syncranial architecture (Text-fig. 38) is essentially of the *H. saxicola* type (Text-fig. 34; see also p. 59 above), but the oral definition is peculiar. There is a high proportion of tricuspid teeth in the outer tooth rows of both jaws, but especially in the lower one. The other teeth in these series are either slender bicuspids or unicuspids; that is, like the teeth of *H. saxicola*. Unlike *H. saxicola*, none of the '*tridens*' group shows any enlargement of the median teeth on the lower pharyngeal bone (and never, as in some *H. saxicola* individuals, an enlargement of the bone itself; see Greenwood, 1960).

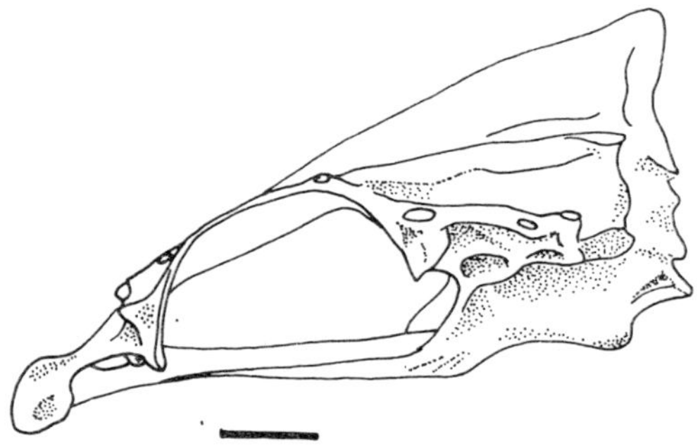

FIG. 38. Typical neurocranial form in a member of the *H. tridens* lineage (*ex H. tridens*). (Scale = 3 mm.)

Neither the '*tridens*' group nor *H. saxicola* exhibits any specialized characters obviously associated with their feeding habits (unless the slightly enlarged median pharyngeal teeth of *H. saxicola* are considered thus; but the known diet for this species does not indicate any increase in the number of molluscs eaten as compared with generalized insectivores). It is, therefore, impossible to consider either the '*tridens*' species or *H. saxicola* as the apomorph (i.e. derived) sister group of the other. That both should be considered part of the same phyletic lineage seems likely from their shared syncranial specializations.

At least one other species, *H. melichrous*, might be included in this trophic group (Greenwood & Gee, 1969). It has a rather generalized dentition of mixed bi- and unicuspid teeth, but the occasional tricuspid does occur in the posterior part of the lower jaw. *Haplochromis melichrous* differs from members of the '*tridens*' group in its neurocranial morphology, and in these characters resembles members of the '*serranus*' group piscivores (Greenwood, 1967). The dentition of *H. melichrous* is sufficiently generalized (despite the occasional tricuspid teeth) for it to be of little value as a phyletic indicator. Skull and jaw characters are also not particularly clear-cut in this context, although they do seem to exclude the possibility of relationship with species in the '*tridens*'–'*saxicola*' lineage. For the moment, *H. melichrous* can only be associated with the '*serranus*' group, to be discussed on p. 81 *et seq*.

In summary, the benthic crustacean-eating group seems to comprise two lineages. The larger one is a derivative of the *H. saxicola* line, the smaller (*H. melichrous* only) is of uncertain affinity but could possibly be related to the '*serranus*' piscivore lineage (Text-fig. 70).

Scale eating species

Comparatively little is known about the anatomy of *H. welcommei* (Greenwood, 1966b) the unique lepidophage in Lake Victoria. The slender, moderately elongate body (Text-fig. 3), and the superficial head shape of this species are not unlike those of species in the '*tridens*' group discussed above. The broad bands of inner teeth in *H. welcommei* (Text-fig. 39) are a specialization not found in that group, and are in fact not approached by the dental pattern in any other species (except, perhaps,

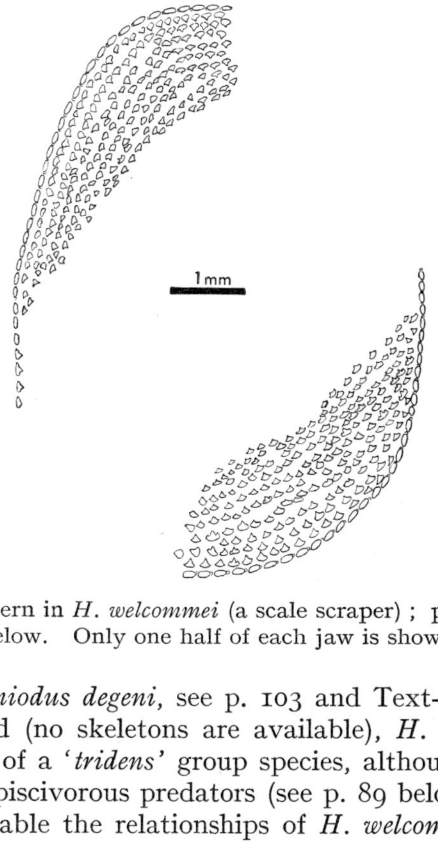

FIG. 39. Jaw tooth pattern in *H. welcommei* (a scale scraper) ; premaxilla above, dentary below. Only one half of each jaw is shown.

the monotypic *Platytaeniodus degeni*, see p. 103 and Text-fig. 73). Judging from radiographs of the head (no skeletons are available), *H. welcommei* does have a neurocranium like that of a '*tridens*' group species, although it also resembles *H. paraguiarti*, one of the piscivorous predators (see p. 89 below). Until good osteological material is available the relationships of *H. welcommei* must remain enigmatic, although the indications are of affinity with the '*tridens*' lineage (Text-fig. 70).

Mollusc eating species

The clear-cut dichotomy in feeding methods among species in this trophic grade has been discussed in detail on p. 37.

Those species that wrench snails from their shells (or crush the shell orally) constitute one of the best defined phyletic lineages within the whole flock. Specializations shared by all four species (*H. sauvagei* [Text-fig. 19], *H. prodromus*, *H. granti* [Text-fig. 20] and *H. xenognathus* [Text-fig. 21]) are the strongly decurved, in some species almost vertical, preorbital skull, a stout lower jaw, the stout unicuspid teeth

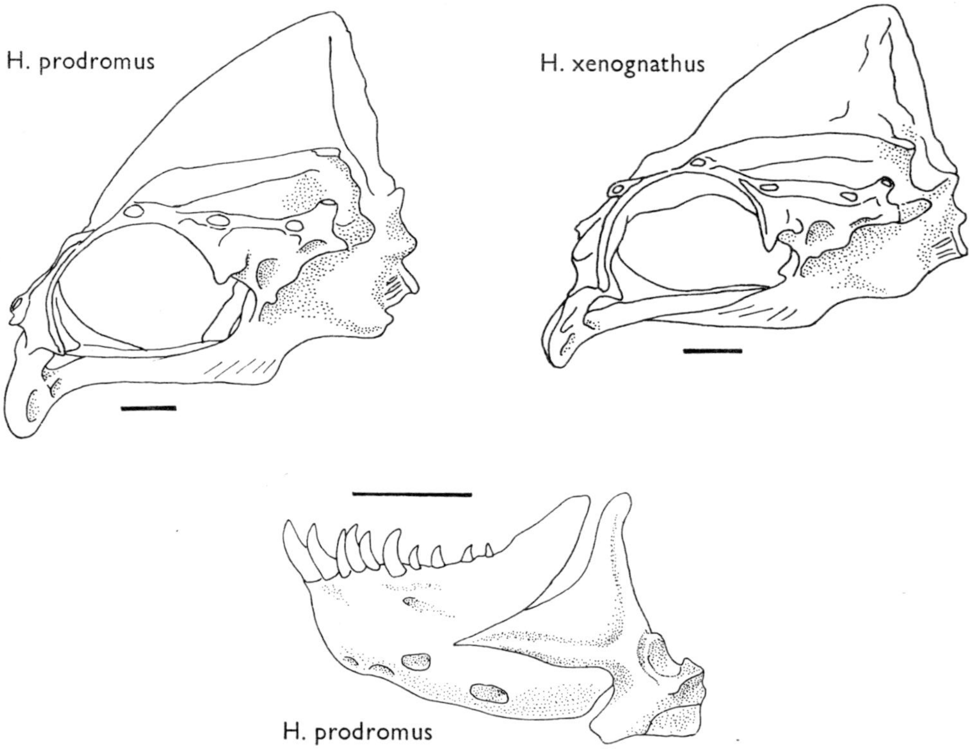

FIG. 40. Neurocranial and jaw form in two species of mollusc-eating *Haplochromis*, both oral-shellers. (Scale = 3 mm.)

with strongly incurved crowns, and the broadly arranged rows of unicuspid inner teeth (see Text-figs 40, 41 and 66 ; also Greenwood, 1957). Smaller members of these species (that is fishes between 80 and 100 mm long, depending on the maximum size reached in the species) do have bicuspid outer teeth, or a mixture of bi- and unicuspids. The bicuspids, like the unicuspids, are strongly recurved, thus contrasting with the barely recurved generalized type of bicuspid. The lower jaw is shorter than the upper in all species but *H. granti*, where it is sometimes a little longer. *Haplochromis granti* also differs from other members of the group in having the mouth inclined upwards at a small angle ; it is horizontal in the others.

Anatomically, *H. xenognathus* is the most specialized species, *H. sauvagei* and *H. prodromus* the least specialized (*cf* Text-figs 40 and 41) ; but these degrees of difference are not apparent in feeding habits, which are identical (Greenwood, 1957). *Haplochromis xenognathus* differs chiefly in having some lower jaw teeth implanted horizontally, the inner tooth rows in much broader bands, with those of the dentary presenting a convex occlusal surface (see Text-fig. 41C). *Haplochromis sauvagei* and *H. prodromus* are virtually identical in their oral, dental and neurocranial morphology, but individuals of the latter species reach a larger size (130, *cf* 105 mm).

The monotypic genus *Macropleurodus bicolor* is, on the basis of its syncranial architecture, a member of the same species group. The jaw teeth of *M. bicolor* (Text-fig. 75) differ from those of other species in being coarser, more strongly incurved and in generally retaining traces of a minor cusp (Text-fig. 75B). This cusp persists as a slight, vertical hump on the near-horizontal occlusal surface presented by the major cusp. The dentigerous arms of the premaxillae are bowed (usually more so on one side than the other) ; consequently one side of the mouth is slightly

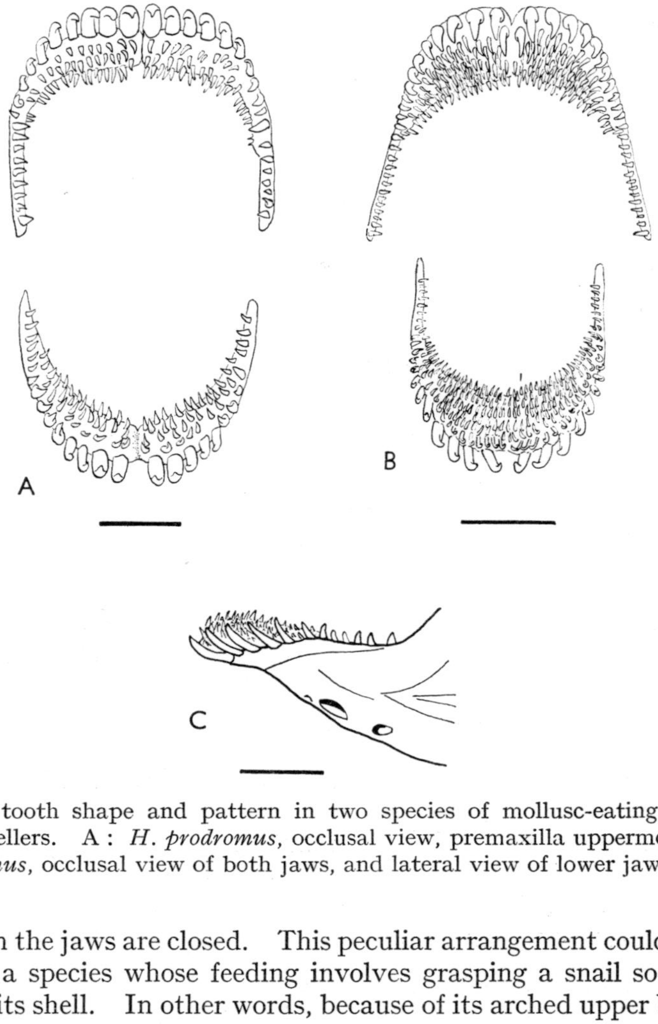

FIG. 41. Jaw tooth shape and pattern in two species of mollusc-eating *Haplochromis*, both oral shellers. A : *H. prodromus*, occlusal view, premaxilla uppermost. B and C : *H. xenognathus*, occlusal view of both jaws, and lateral view of lower jaw respectively.

open even when the jaws are closed. This peculiar arrangement could be of adaptive significance in a species whose feeding involves grasping a snail so that it cannot withdraw into its shell. In other words, because of its arched upper border the gape is increased without an increase in the extent to which the dentary would otherwise have to be lowered.

The near vertical ethmoid region (over which the ascending processes of the premaxillae run) in all *H. sauvagei*-group species (see Text-figs 40 and 66) produces an unusual jaw opening movement, namely downward and slightly forward rather than mostly forward ; the short and deep adductor mandibulae muscles ensure a powerful and rapid closing action for the dentary. The strong jaws and teeth, coupled with these jaw movements, all indicate a high level of adaptation for the unusual feeding methods practised by members of the lineage. It is interesting to see that the ethmo-vomerine bloc in the skull of these fishes is slightly longer than the near-horizontal ethmo-vomer of a generalized skull. (In fact, it is relatively as long as in many piscivorous species groups ; see pp. 80–93.) Presumably the lengthened ethmo-vomer gives added support and backing to the premaxillary processes when the upper jaw is protruded downwards and the fish is wrenching a snail from its shell.

A possible phyletic relationship between the *H. sauvagei* group and the insectivorous species *H. chilotes* and *H. chromogynos* has already been suggested (p. 61). The skull in these latter species is rather less strongly decurved (see Text-figs 32, 40 and 66), and this coupled with their relatively unspecialized feeding habits (except, on occasion, in *H. chilotes*, see p. 31), leads me to rank the *H. sauvagei* lineage as the apomorph sister group of the *H. chilotes–H. chromogynos* pair. Phylogenetically it may be significant that black-and-white piebald female polymorphs (see p. 53) are recorded in four of the six species in this lineage (*H. sauvagei*, *Macropleurodus bicolor*, *H. chilotes* and *H. chromogynos* [where it is, indeed, the 'normal' female coloration]). Similar polychromatism occurs in several other species (none closely related to the '*sauvagei*' lineage) but never elsewhere is it known from so many species in a presumed lineage (see pp. 53–54 above).

When more is known about certain Lake Victoria species, for example *H. crassilabris* (see Regan, 1922), the *H. sauvagei* lineage may have to be expanded. Hopefully it will then prove possible to find the group's plesiomorph relatives.

The second mollusc eating group is made up from those species in which the prey is crushed between the upper and lower pharyngeal teeth and bones.

Two species, *H. ishmaeli* and *H. pharyngomylus* (Greenwood, 1960 ; also p. 37 above), have the most highly developed pharyngeal mills (see Text-fig. 5A). The lower bone is greatly enlarged and most of its teeth are molariform ; the upper bones are correspondingly modified. The skull is essentially like that of the generalized type, except for the better-developed ventral apophysis on which the upper pharyngeal bones articulate, see Text-figs 43 and 44. This apophysis is formed, as in all *Haplochromis* species, mainly from the parasphenoid and basioccipital bones, but there is also a small contribution from the prootics as well (Text-fig. 44D). The jaws and oral dentition are essentially those of a generalized species like *H. pallidus* (see Text-figs 33 and 43). What differences there are lie within the limits of variation resulting from allometric growth ; individuals of *H. ishmaeli* and *H. pharyngomylus* reach a much larger size (125–135 mm) than do those of the '*pallidus*' insectivore group (*ca* 100–105 mm).

Superficially, *H. obtusidens* closely resembles *H. ishmaeli* and *H. pharyngomylus* (see Greenwood, 1960), but its pharyngeal mill is at an intermediate level of hypertrophy, the lower bone moderately stout and its dentition partly molarized ; see Text-fig. 44B. The paryngeal apophysis on the skull is also at an intermediate level of specialization.

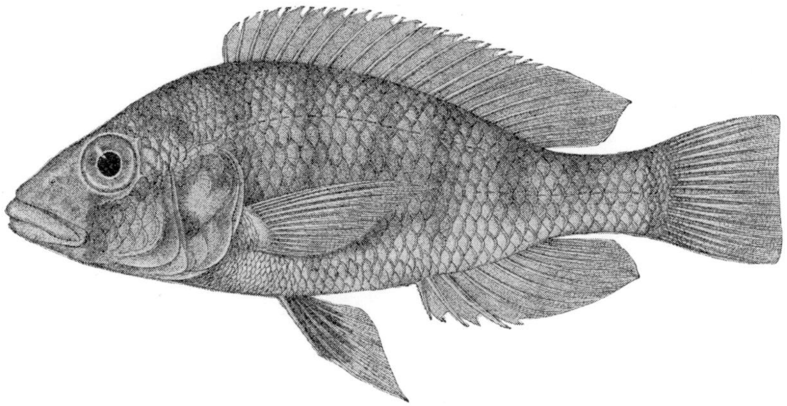

FIG. 42. *Haplochromis riponianus.* (Natural size.)

The skull of *H. obtusidens* taken as an entity is, however, more like that of *H. empodisma* (*cf* Text-figs 32 and 43), an insectivorous, bottom-detritus eating species (see p. 63 above). Thus, although I have previously indicated a close relationship between *H. obtusidens* and the *H. ishmaeli* pair (Greenwood, 1960) and therefore a possible phyletic lineage of increasing specialization, I now suspect that this was an error. Recent analysis of various character complexes suggests that although *H. empodisma* and *H. obtusidens* are related (and are members of the '*empodisma*' lineage, see p. 63) and although *H. ishmaeli* and *H. pharyngomylus* are also interrelated, the latter species pair belong to a distinct lineage (Text-fig. 70).

The most likely plesiomorph sister species for *H. ishmaeli* and *H. pharyngomylus* is *H. humilior* (Text-figs 5C and 43; see also Greenwood, 1960). In *H. humilior* the pharyngeal mill is moderately developed (and shows a high degree of intraspecific variability, see Greenwood, *op. cit.*) and the pharyngeal apophysis shows some departure towards the '*ishmaeli*' condition (an increased area of basioccipital in the facet; see Text-fig. 44C). But the overall form of the neurocranium in *H. humilior* is of the basic insectivore type, with a short and decurved preorbital face (see Text-fig. 43). Apart from the larger pharyngeal apophysis, the neurocranium in *H. ishmaeli* is identical with that of *H. humilior* (Text-fig. 43).

Haplochromis humilior could well be derived from a species like *H. pallidus*, a taxon in which there is already some slight hypertrophy of the lower pharyngeal teeth (see p. 59). The close correlation that can exist between the relative massiveness of the pharyngeal bones (and their teeth) and the degree to which the neurocranial apophysis is developed has been clearly demonstrated in both natural and experimental populations of the mollusc-crushing *Astatoreochromis alluaudi* (Greenwood, 1959a, 1965c). It is also apparent in ontogenetic series of *H. ishmaeli* and *H. pharyngomylus* (personal observations). Thus the interspecific differences seen in that region of the skull in *H. pallidus* and *H. humilior* on the one hand and *H. ishmaeli* and *H. pharyngomylus* on the other would seem to be associated with increasing hypertrophy of the pharyngeal mill.

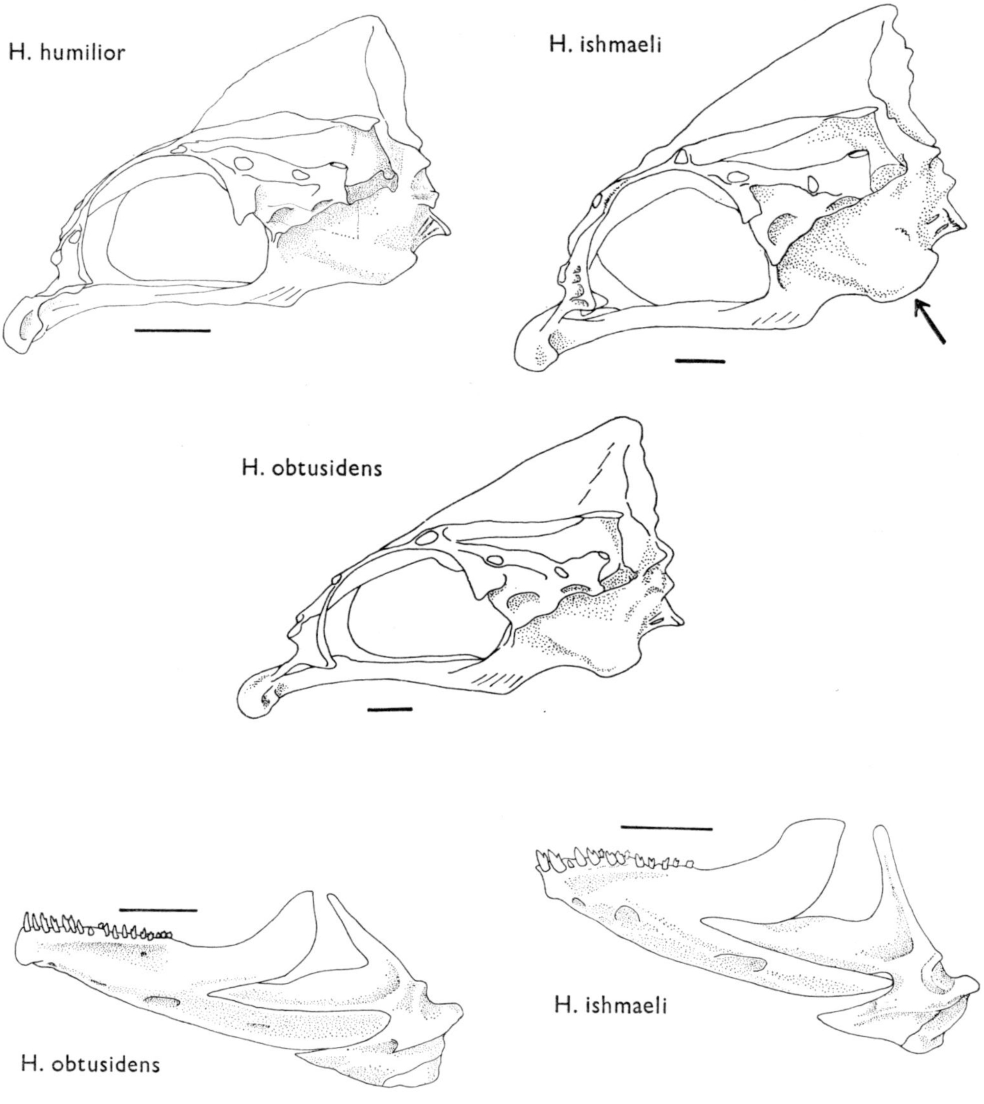

FIG. 43. Neurocranial and jaw form in some mollusc-eating species (all pharyngeal crushers). Arrow indicates the apophysis (formed from the parasphenoid and basi-occipital bones) on which the upper pharyngeal bones articulate (see Text-fig. 44). (Scale = 3 mm.)

All the species considered so far are those in which Mollusca (especially gastropods) are the chief food source. Two other species, *H. riponianus* and *H. theliodon* (Greenwood, 1960), are mixed insect–mollusc eaters.

Haplochromis riponianus shares the syncranial (Text-fig. 34) and dental characteristics of *H. saxicola* but with a more consistent hypertrophy of the median tooth rows on the lower pharyngeal bone, and generally some enlargement of the bone as well. As in all other species with similar characteristics of the pharyngeal mill, the degree of bone and tooth hypertrophy is size correlated. Phylogenetically, *H. riponianus* is clearly the apomorph sister species of *H. saxicola*.

Lack of sufficient material for anatomical studies makes it impossible to suggest the relationships of *H. theliodon* (see discussion in Greenwood, 1960).

In brief, at least four distinct lineages are involved in the mollusc-eating adaptive radiation ; each can be traced back to an essentially insectivorous level still represented by different species in the flock (Text-figs 65 (3), 66, 68 (2) and 70).

Predators on larval and embryo cichlid fishes

This, the paedophage radiation, was described at length on pp. 31–37. There is a clear-cut diphyletism in the group, the branches comprising *H. obesus*, *H.*

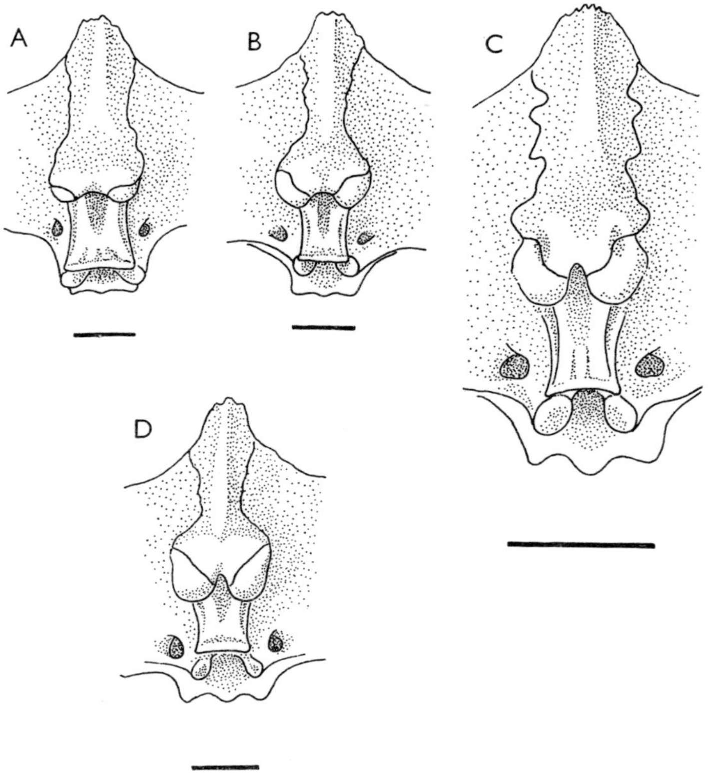

FIG. 44. Ventral view of the neurocranial apophysis for the upper pharyngeal bones, showing the relative increase in the size of the basioccipital facets correlated with increasing stoutness of the pharyngeal bones and teeth (see Text-fig. 5). (Scale = 3 mm.)
A : *H. empodisma* (typical of condition found in most species).
B : *H. obtusidens* (a species with slightly enlarged pharyngeal bones and teeth).
C : *H. humilior* (with moderately enlarged pharyngeal bones and teeth).
D : *H. ishmaeli* (massive pharyngeal bones and teeth).

maxillaris and *H. melanopterus* (Text-figs 3 and 17), and *H. cryptodon, H. microdon* and *H. parvidens* (Text-figs 3 and 18) respectively (Greenwood, 1959b). There are two other paedophagous species, *H. cronus* (Greenwood, *op. cit.*) and *H. barbarae* (Greenwood, 1967) whose lack of anatomical specializations relative to the main lineages make their relationships difficult to determine (Text-figs 16 and 29).

In both lineages the anatomical specializations involve, particularly, development of a large mouth that is both protrusible and laterally distensible, a reduction in the number of jaw teeth, together with their restriction to the anterolateral parts of the

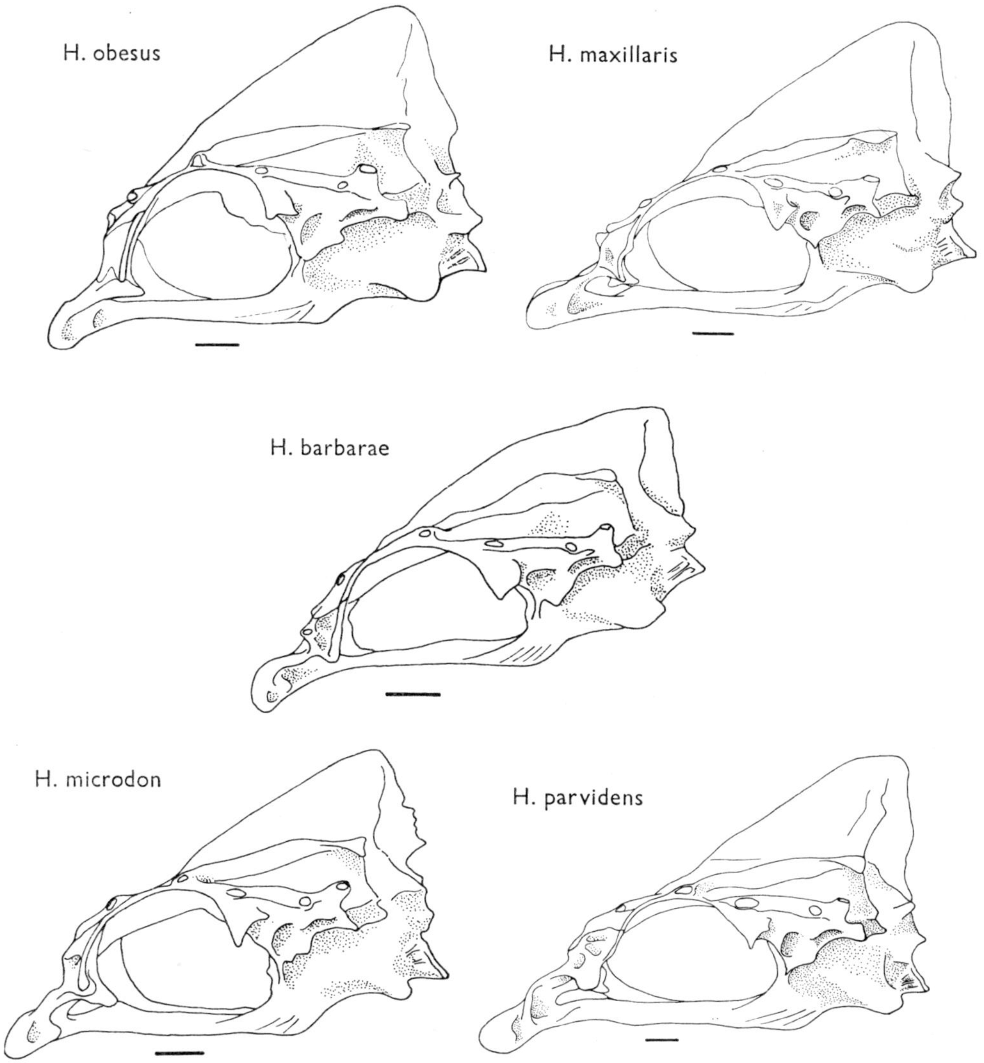

FIG. 45. Neurocranial form in paedophagous species (see text p. 75 ; also Text-fig. 46). (Scale = 3 mm.)

jaw bones, and their almost complete burial in the thickened oral mucosa. (In one species, *H. maxillaris* the teeth are further covered by an inward curvature of the upper lip tissue.)

The functional anatomy of the jaws in these species has not been adequately analysed. Dissection does not reveal any marked departure from a typical arrangement and interrelationship of the jaw bones, ligaments and muscles. However, in all species the ascending premaxillary processes are long, the maxilla is rather bullate distally, and in all except *H. obesus* and *H. maxillaris*, the lower jaw shows a marked narrowing at about the middle of its length. As a result of this latter peculiarity, the lower jaw closes within the upper jaw. Gauged from the effects of manipulation on freshly dead fishes, the mouth is more protrusible, and there is greater lateral displacement of the upper jaw in members of the *H. cryptodon* lineage than in those of the *H. obesus* line.

Haplochromis obesus, *H. maxillaris* and *H. melanopterus* are characterized by all or most of the outer dentary teeth having the cusp inclined outwards (Text-fig. 46A). This tooth shape is found in no other Lake Victoria *Haplochromis* species. *Haplochromis obesus* has a neurocranium rather like the generalized type (Text-fig. 45). That of *H. maxillaris* (Text-fig. 45) has a slightly straighter preorbital face, in this respect approaching the *H. empodisma* skull type (see Text-fig. 32). Despite this difference in neurocranial shape, the shared peculiarity in lower jaw tooth shape seems to exclude the possibility of there being closer relationships between *H. empodisma* and *H. maxillaris* than between the latter species and *H. obesus* or *H. melanopterus*. Judging from the relative extent of jaw protrusibility in *H. maxillaris* (possibly correlated with the flatter skull profile) this species is more specialized than *H. obesus*. Regrettably no skeletal material of *H. melanopterus* is available, and the species is included in the *H. obesus* lineage mainly on the characteristic shape of its lower jaw teeth (see Greenwood, 1959b).

Neurocranial shape in the *H. cryptodon* and *H. parvidens* lineage (Text-figs 45 and 68) is essentially of the type found in the less extreme members of the '*prognathus*' group of piscivorous predators (see Greenwood, 1967, and pp. 85–89 below). *Haplochromis parvidens* (Text-fig. 45) has the most '*prognathus*'-like skull, *H. microdon* and *H. cryptodon* the least modified in that direction (Text-fig. 45). Dentary shape is correlated with neurocranial shape, *H. parvidens* having the most extreme degree of anterior narrowing (Text-fig. 46C). It is this unusual and unique shape of the dentary, combined with overall neurocranial morphology, that provide the principal evidence for the presumed monophyly of the group. Morphologically, the oral teeth in these fishes are like those of typical piscivorous predators in the '*serranus*' and '*prognathus*' groups (see Greenwood, 1960, 1967; and pp. 82–89 below).

The maximum adult size reached by members of both paedophage lineages is in the range of 130–170 mm; that is, these are among some of the larger *Haplochromis* species in Lake Victoria.

Haplochromis barbarae (Greenwood, 1967) and *H. cronus* (Greenwood, 1959b) are included as members of the paedophage group solely on the basis of their feeding habits. Neither species shows any of the dental or oral specializations of either the *H. obesus* or *H. cryptodon* lineages.

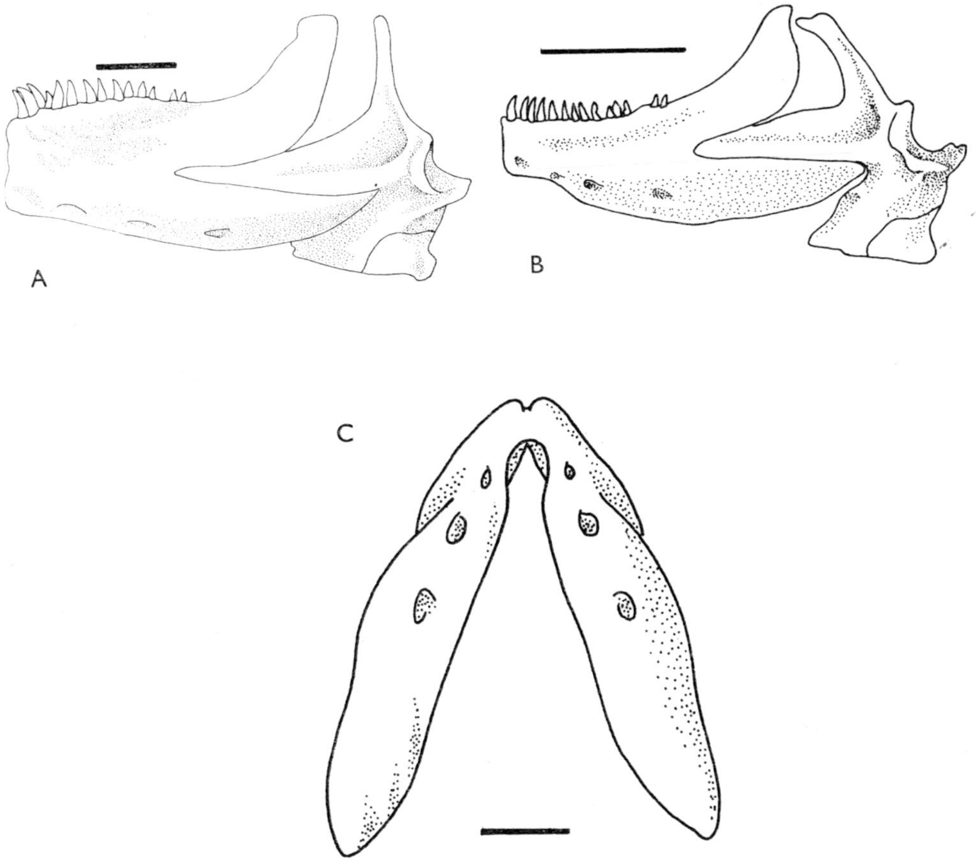

FIG. 46. Lower jaw in two paedophagous species. A : *H. obesus*. B and C : *H. parvidens*. A and B in lateral view, C in ventral view to show characteristic outline of the dentary in *H. parvidens*-group paedophages. (Scale = 3 mm.)

In *H. cronus* the outer jaw teeth are moderately stout unicuspids, not noticeably reduced in number or size, nor restricted in their distribution nor deeply embedded in the mucosa. Little is known about the syncranial architecture of the species, except from partial dissection of one fish. Judging from information gleaned in this way, *H. cronus* does not have the syncranial characteristics of any piscivorous group. Like *H. ishmaeli*, which it resembles in many ways, *H. cronus* seems to share the syncranial morphology of a generalized insectivore, but it is larger than any of these fishes (adult sizes range to at least 135 mm). Unlike the majority of Lake Victoria species, irrespective of their trophic associations, *H. cronus* has almost the entire surface of the caudal fin densely covered with small scales (Greenwood, 1959b).

As *H. cronus* seems to have the syncranial architecture and dentition of a generalized *Haplochromis*, it could be ranked as the plesiomorph sister group to the *H.*

obesus lineage. These fishes, more than those of the *H. cryptodon* lineage, have retained a near-generalized skull and jaw morphology (despite the slight but obvious specializations of the latter). For these reasons, the relationships of the *H. obesus* lineage cannot be defined readily. It could be considered either an independent offshoot from the basal complex of insectivorous species (p. 63 above) or as a derivative of the *H. cinereus–H. squamulatus* line.

Haplochromis barbarae, on the other hand, has the syncranial architecture (Text-fig. 45) of the '*altigenis*' predator group (see p. 82), and the type of oral dentition found in small individuals of that species assemblage (adult *H. barbarae* are 85–106 mm long). Basically, skull morphology in members of the '*altigenis*' group is not greatly different from that in less specialized species of the '*prognathus*' group (Greenwood, 1962, 1967 ; also pp. 82 & 85 below). It will be recalled that the neurocranial form in the *H. cryptodon* lineage is also like that among basal members of the '*prognathus*' group (see above, p. 77). Thus, in view of its feeding habits and of its anatomical specializations, *H. barbarae* could be included in the *H. cryptodon* lineage as the plesiomorph sister group of the other species combined.

Fryer & Iles (1972) believe that there is '. . . lack of striking diagnostic indications of feeding habits among these species . . .' and that this '. . . is not really surprising'. They go on to argue that the collection of such 'soft morsels' as fish embryos and larvae does not require the evolution of a specialized anatomy in the predator. Furthermore, Fryer & Iles believe the morphological variation found in paedophage species suggests '. . . dissociation of structure and function . . .'. I fail to find any evidence supporting a single one of these ideas.

In both paedophage lineages there is ample evidence for anatomical differentiation in jaw morphology, especially in the *H. cryptodon* lineage (see above), and in the buried dention of all species (a specialization that Fryer & Iles themselves acknowledge). Finally, the manifestation of these various anatomical traits is very obvious to anyone who has handled fresh specimens and compared the extent of their jaw protrusion with that in most other Lake Victoria *Haplochromis* species. (Some predator species are excluded from this generalization, but these belong to lineages whose basic cranial specializations are greater than those of the paedophages.) As for the postulated dissociation of structure and function, surely the anatomical evidence negates any such idea in its entirety?

I find it surprising that Fryer & Iles should think that there are no diagnostic indications of feeding habits among the paedophages. Once one has associated the morphology of any paedophage species with its diet, it is remarkably easy to recognize any other paedophage at sight, or so has been my experience and that of others working in the field, even when dealing with the species of another lake.

The intraspecific variability seen in paedophagous species (Greenwood, 1959b) is high, but no higher than that in many other species. It is perhaps the generally bizarre cranial morphology of these fishes that accentuates this variability. I can find nothing to suggest, as Fryer & Iles imply, that paedophage anatomy is not directly correlated with feeding habits (i.e. is non-adaptive) and that there has been a consequent relaxation of selection pressure leading to greater deviation from the mode in these characters. Fryer & Iles' views, I suspect are coloured by their belief in the idea that paedophages do not actively obtain their prey from a parent fish, but snap up voluntarily jettisoned young and embryos (see p. 33 above).

I find myself more in agreement with Fryer & Iles (*op. cit.*) when it comes to the question of how paedophagous habits originated. The habit probably stems from a facultative response to the appearance of young liberated by the parent during the normal course of brood care. That many *Haplochromis* species, irrespective of their usual feeding habits, respond to the sudden appearance of small objects in the water is often demonstrated in Lake Victoria. On several occasions at Nasu Point beach near Jinja, every species present was found to be gorged on termites after a heavy hatch of these insects had been carried into the lake. On other occasions all species (including specialized piscivores) had ingested large quantities of colonial blue-green algae that were floating in the water after a period of rough weather (unpublished personal observations).

Whatever the origins of paedophagy, there are now two phyletic lines of paedophagous species in Lake Victoria (the only lake, apart from Lakes Edward and George, in which the habit has evolved [Greenwood, 1973a]). One lineage (*H. obesus* and related species) evolved from a basic piscivore stem, the other (*H. cryptodon* and allies) from a derived, piscivore line (Text-figs 67, 68 and 70).

Piscivorous predators

The 35 or more species in this trophic group present the most complicated phyletic puzzle in the whole flock. In part this is attributable to the larger number of species involved, but mainly it is because of the few characters that can be used to determine relationships. Whereas in other groups oral and pharyngeal dentition provide indications of relationship, among predators the dentition is essentially a uniform one of strong, somewhat recurved, unicuspids (Text-figs 4B and 57). Even small individuals have this type of dentition, although in juveniles the teeth are bicuspid and of the generalized type. Body form, too, is of little value. The only character complex seemingly of phyletic value is the syncranium, especially the shape of the neurocranium (Greenwood, 1962, 1967).

Difficulties are also encountered when interpreting these neurocranial features, and it seems unlikely that a satisfactory scheme of phyletic interrelationships can be achieved. To start with, it is impossible to tell whether the main groups of predators are of monophyletic origin, even if the concept of monophylogeny is broadened to include origin from more than one species provided the species are themselves sister taxa.

The chief diagnostic features of the piscivorous predator grade concern adaptations for feeding on larger and faster moving objects than are utilized as prey by the other trophic groups. This has involved elongation of the preotic part of the skull, elongation of the jaws (often associated with a marked upward inclination of the gape), and the development of a strong, unicuspid dentition adapted for holding a struggling prey fish. In short, the production of a large-mouthed, streamlined fish of a somewhat greater size than the members of species in other trophic groups.

In previous papers (Greenwood, 1962, 1967) three principal groups of piscivorous predators were defined, chiefly on the grounds of neurocranial shape and proportions.

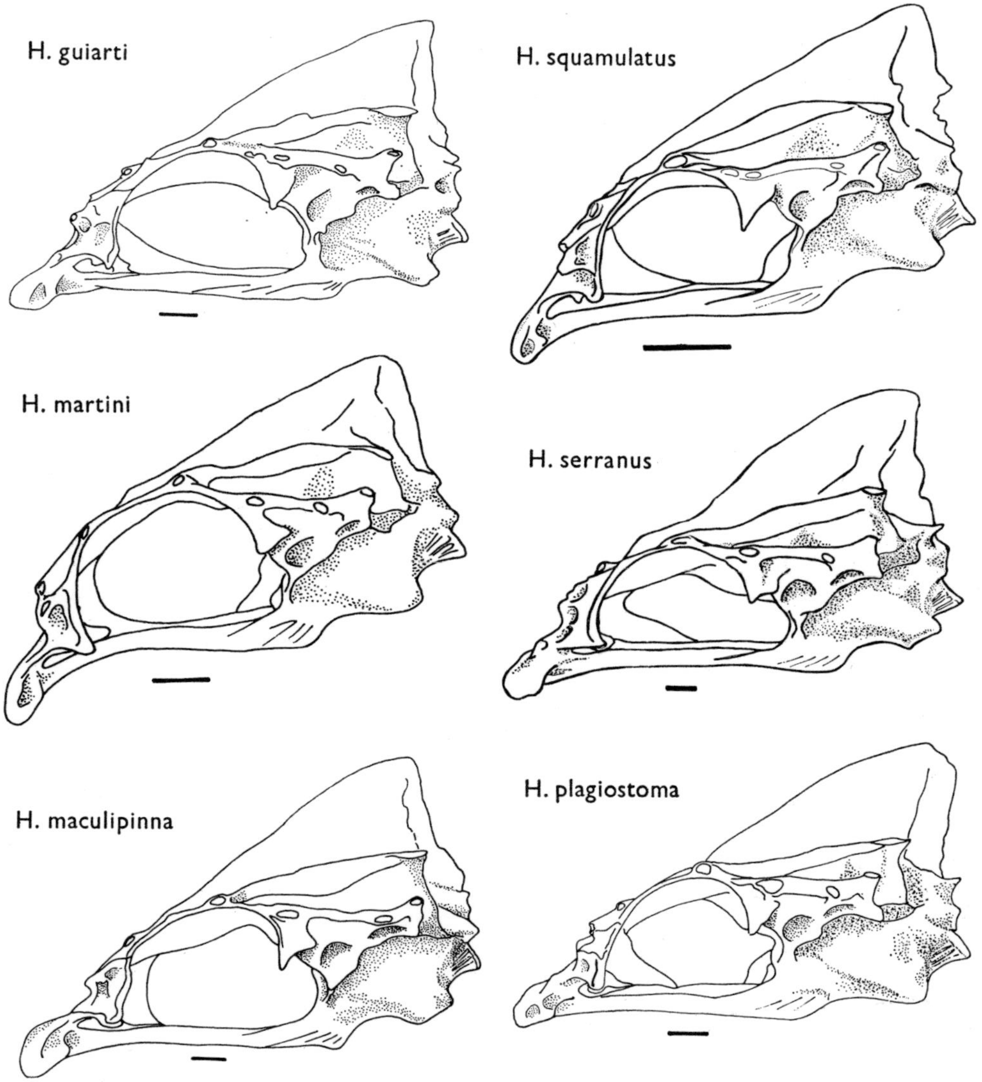

FIG. 47. Neurocranial form in piscivorous predators. The species represented are from the *H. guiarti–H. squamulatus* and the *H. serranus* lineages. (Scale = 3 mm.)

The first or '*serranus*' group (comprising eight species) has a neurocranial form nearest that of the generalized type, but one already showing some elongation of the preotic skull and a marked flattening of the dorsal outline to the preorbital face (Text-figs 47 and 57). The second or '*altigenis*' group (also of eight species) has a more elongate neurocranial outline, with the preotic face slightly longer and sloping less steeply than in the '*serranus*' type (Text-figs 51 and 57). The third or '*prognathus*' group (11 species, Text-figs 51, 57 and 62) shows further accentuation of the '*altigenis*' trend, together with an overall narrowing of the skull. A number of smaller and probably polyphyletic groups, all of uncertain affinities with the major ones, were also recognized (Greenwood, *op. cit.*).

Originally I thought the three major groups were each of monophyletic origin, but with a strong possibility of the '*altigenis*' and '*prognathus*' groups stemming from an ancestor not shared with the '*serranus*' group. However, species discovered recently (Greenwood & Gee, 1969) and further study of the known species have caused me to modify somewhat the views expressed before.

To begin with, it seems that the ancestral skull type is not of the kind found in *H. guiarti* and *H. brownae* (*pace* Greenwood, 1962). These species have a skull form very like the generalized insectivore type (see pp. 57–59), whereas the basic predator skull type ('*serranus*' group) is, in fact, more specialized and akin to that found in *H. saxicola* (see p. 59). *Haplochromis saxicola* is, of course, an insectivore but its whole level of syncranial organization departs from the generalized *H. pallidus* type toward that of the piscivorous predators (including preotic elongation of the neurocranium, marked flattening of the dorsal preorbital profile and a preponderance of uni- over bicuspid teeth). Individuals of *H. saxicola* also reach a larger adult size than do those of the *H. macrops*-like species ; but adult size may be of secondary importance because *H. brownae*, an insectivore–piscivore is no larger than *H. macrops*. The possible relationships of *H. brownae* and *H. guiarti* will be discussed later (p. 85).

Since *H. saxicola* (despite its insectivorous habits) has a neurocranial form already specialized towards that of the piscivorous predators, it seems more reasonable to consider some *H. saxicola*-like species as the ancestor of almost the entire piscivorous predator radiation.

Of the three predator groups originally defined, I would now consider the '*altigenis*' and '*prognathus*' lines to represent a single phyletic assemblage (henceforth called the *prognathus* group). In other words, species of the '*prognathus*' group could be derived both from species at an '*altigenis*' level of specialization and from species that had already reached the '*prognathus*' level. This conclusion was reached after reexamination of the two 'lineages' had shown not only their great similarity, but also the impossibility of demonstrating, unequivocally, that a species in the more specialized lineage could only be derived from an equally specialized taxon rather than as a somewhat more apomorphic derivative of a species at the '*altigenis*' level.

The phyletic integrity of the '*serranus*' group, on the other hand, remains unchanged, although its membership has to be increased (see p. 84 below). There is no evidence to suggest that any member of the '*prognathus*' lineage might have been derived from a '*serranus*'-like species. The specialized neurocranial form in the '*prognathus*' lineage is more readily derived from that of the presumed *H. saxicola*-like common ancestor of the two lineages than through a '*serranus*'-like form. Neurocranial shape in the '*serranus*' lineage is less variable interspecifically than it is in its apomorph sister group ('*prognathus*'), and the general facies of its constituent species is also more uniform (*cf* Text-fig. 68 (3) with Text-figs 68(4) and 69).

Determining the intrarelationships of species within the two lineages is very difficult, and the results presented here must be considered extremely tentative.

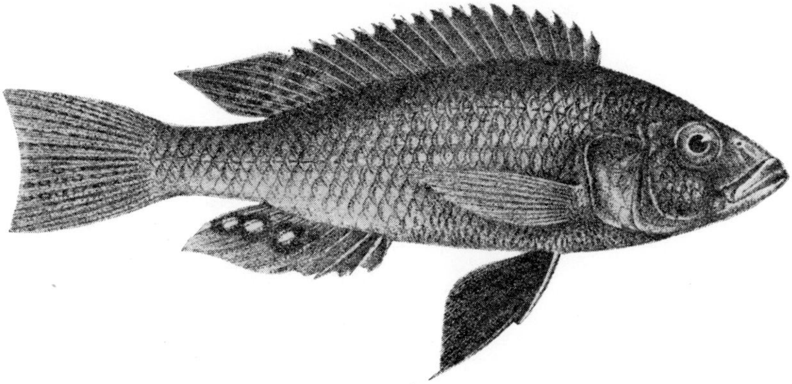

FIG. 48. *Haplochromis guiarti*. (About $\frac{2}{3}$ natural size.)

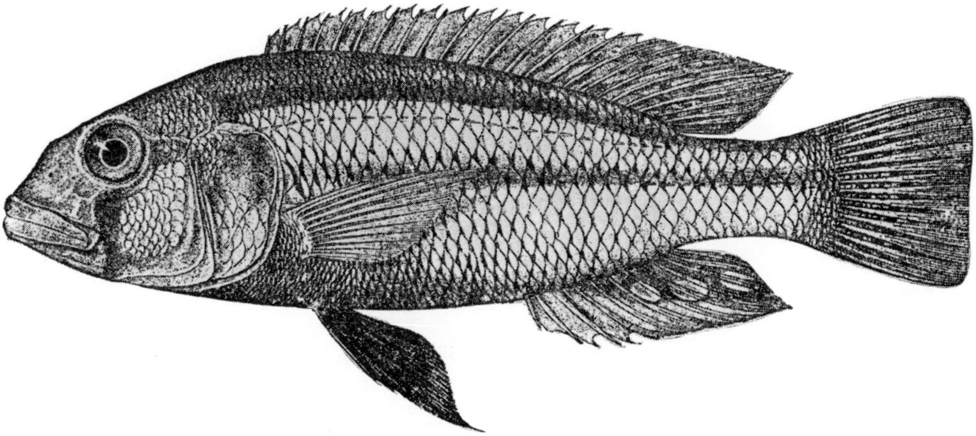

FIG. 49. *Haplochromis squamulatus*. (About half natural size.)

FIG. 50. *Haplochromis martini*. (About natural size.)

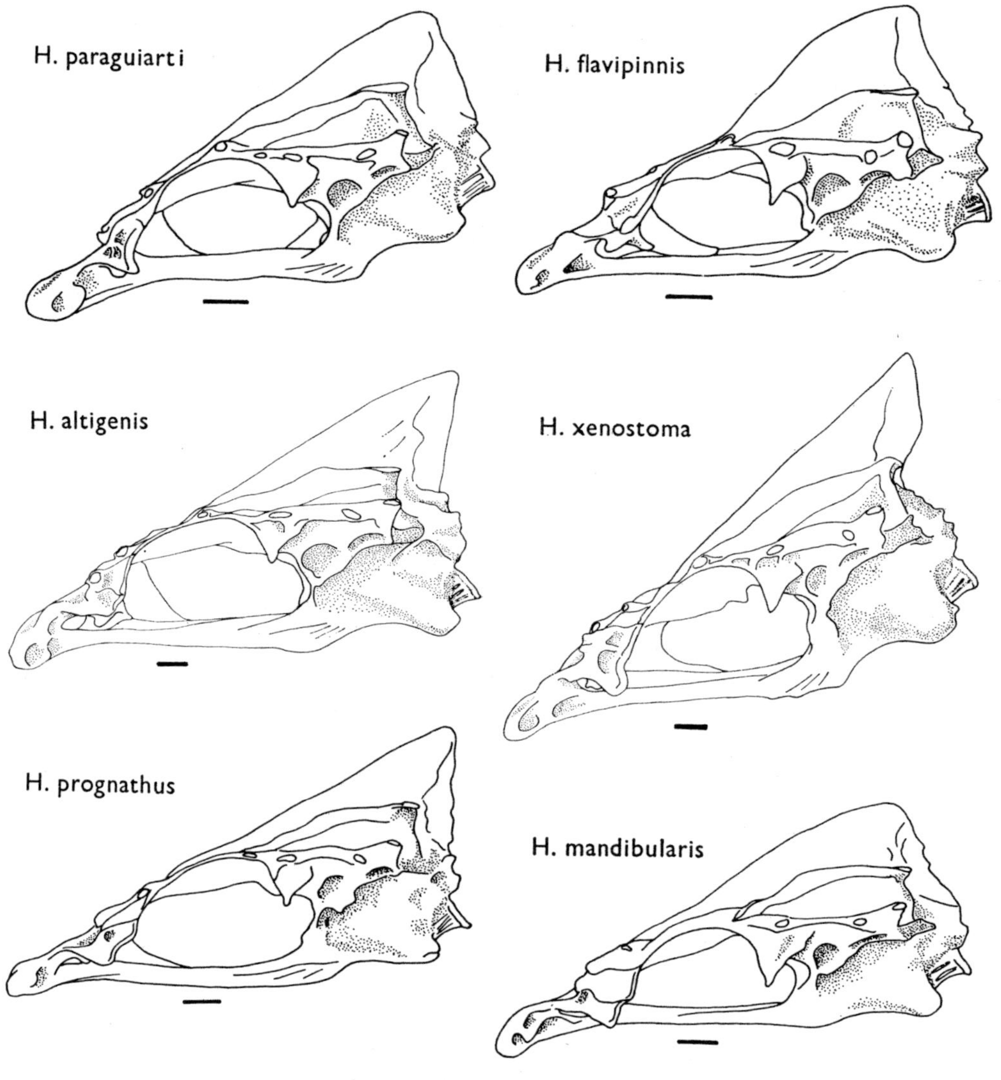

FIG. 51. Neurocranial form in piscivorous predators. (Scale = 3 mm.)

To the eight species originally placed in the '*serranus*' group (*H. serranus* [Text-fig. 11], *H. victorianus*, *H. nyanzae*, *H. speki* [Text-fig. 12], *H. maculipinna* [Text-fig. 47], *H. boops*, *H. thuragnathus* and *H. pachycephalus* ; see Greenwood, 1967), three others should be added, *viz H. cavifrons* [Text-fig. 3], *H. plagiostoma* [Text-fig. 47] and *H. decticostoma* (see Greenwood, 1967 ; Greenwood & Gee, 1969). Neurocranial morphology in all three species (Text-figs 47 and 57) is typically that of the '*serranus*' lineage, despite the rather atypical general facies of *H. cavifrons* and *H. plagiostoma* (Text-fig. 3). Indeed, apart from these two species the '*serranus*' lineage has a remarkably uniform facies. *Haplochromis cavifrons* and *H. plagiostoma* not only differ in their gross appearance (from one another as well as from other species) but also have coloration that is outstandingly different.

Because of the few noticeably specialized features shown by its members, intra-lineage relationships are not at all distinct. Once again *H. cavifrons* and *H. plagiostoma* are outstanding (because of their oblique jaws), but there is little else to indicate a particularly close relationship between them. Certainly their coloration would belie any such suggestion (see Greenwood, 1962).

As mentioned earlier, *Haplochromis guiarti* (Text-fig. 48), formerly considered a member of the '*altigenis*' group (see Greenwood, 1967), has a neurocranial shape little different from that of the generalized *H. pallidus* type (see Text-fig. 47 ; and pp. 58–59 above). What specialization there is, is manifest in the somewhat more elongate preotic skull, and the slightly less curved preorbital skull roof. The dentition in *H. guiarti* is, however, predominantly unicuspid in fishes of a size that in *H. pallidus*-like species would have bicuspid teeth only. In this respect *H. guiarti* is a more specialized (and larger) form of *H. brownae* (Greenwood, 1962), a species little removed in neurocranial form and in dentition from *H. pallidus.* Thus, *H. guiarti* and *H. brownae* could be sister species, apomorph derivates from the *H. pallidus–H. macrops* species complex.

Three other piscivorous (or predominantly piscivorous) species, *H. michaeli, H. squamulatus* (Text-fig. 49) and *H. martini* (Text-fig. 50), have a neurocranial shape close to that of *H. guiarti* (see Text-figs 47 and 67 ; also Greenwood, 1962, 1967, 1960 for the species respectively). In *H. martini* (maximum length 104 mm) the outer series of jaw teeth are mostly bicuspids, with a few unicuspids intercalated ; in *H. squamulatus* (a larger fish, maximum size 198 mm) a similar admixture of teeth is found in fishes less than 115 mm long, but unicuspids predominate in larger fishes, while in *H. michaeli* (maximum size 145 mm, smallest known specimen 117 m) all the outer teeth are unicuspid.

All three species have very small scales on the pectoral region (a character rarely encountered in the Lake Victoria species), and a well-developed, apparently invariable, midlateral stripe from the opercular margin to the caudal fin origin or slightly beyond (another unusual feature for this species flock). Furthermore, both *H. martini* and *H. squamulatus* have a distinctly yellow ground coloration, also an uncommon feature. (The live colours of *H. michaeli* are not known.)

Although neurocranial shape (Text-fig. 47) in these species (and in *H. guiarti*) is more like the generalized than the '*serranus*' type, it is nevertheless a derived form (see above). Thus *H. guiarti* and the species discussed above may represent a true phyletic lineage, with *H. squamulatus, H. martini* and *H. michaeli* more closely related to one another than any one is to *H. guiarti* (Text-figs 67 and 70). The origin of this presumed lineage, unlike that of the '*serranus*' and '*prognathus*' lineages, is probably from an *H. macrops*- or *H. pallidus*-like species, for example a species akin to *H. brownae* which is part insectivore and part piscivore (Greenwood, 1962).

The 20 species comprising the '*prognathus*' lineage (i.e. the combined '*prognathus–altigenis*' groups of Greenwood, 1962 and 1967 ; see p. 82 above) all have a more

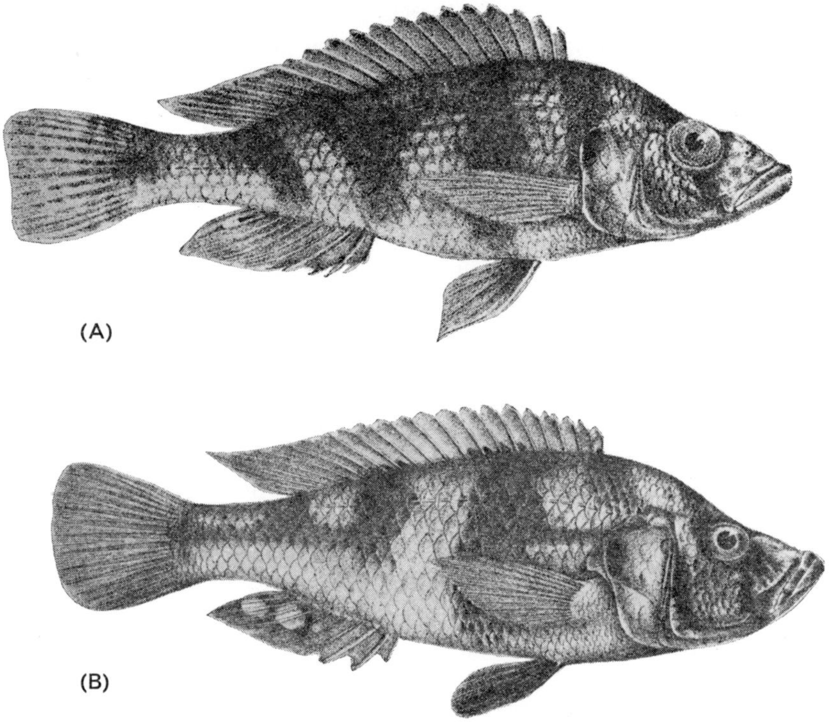

FIG. 52. A : *H. percoides.* (About natural size.) B : *H. flavipinnis.* (About half natural size.)

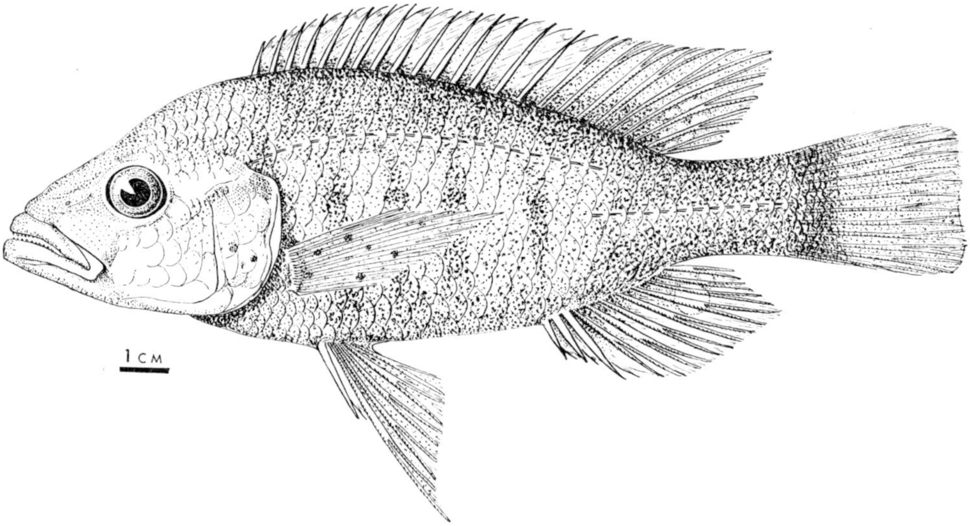

FIG. 53. *Haplochromis paraguiarti.*

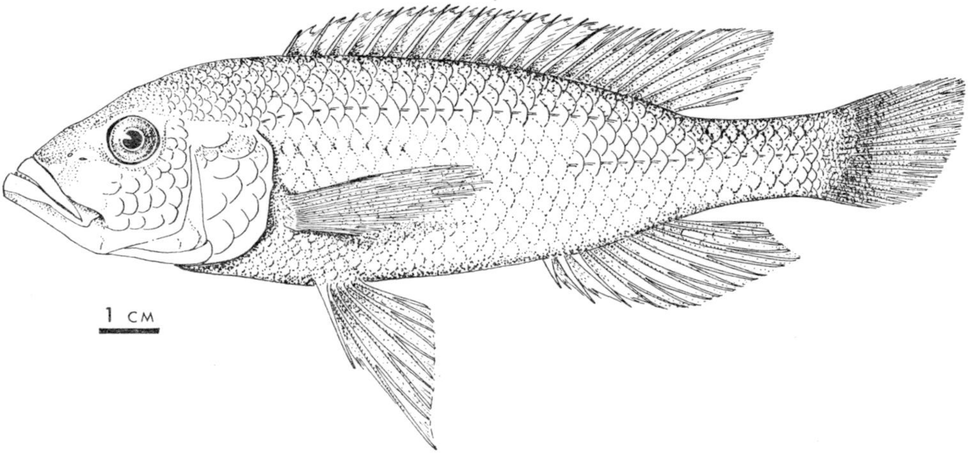

FIG. 54. *Haplochromis pseudopellegrini.*

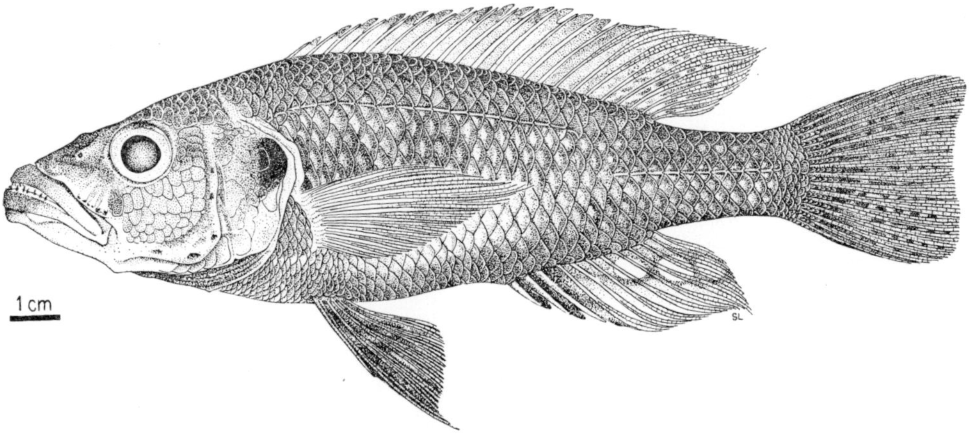

FIG. 55. *Haplochromis gilberti.*

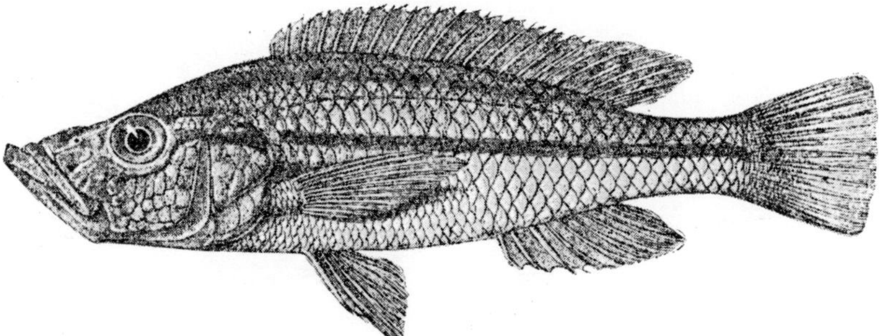

FIG. 56. *Haplochromis xenostoma*; a juvenile fish. (About half natural size.)

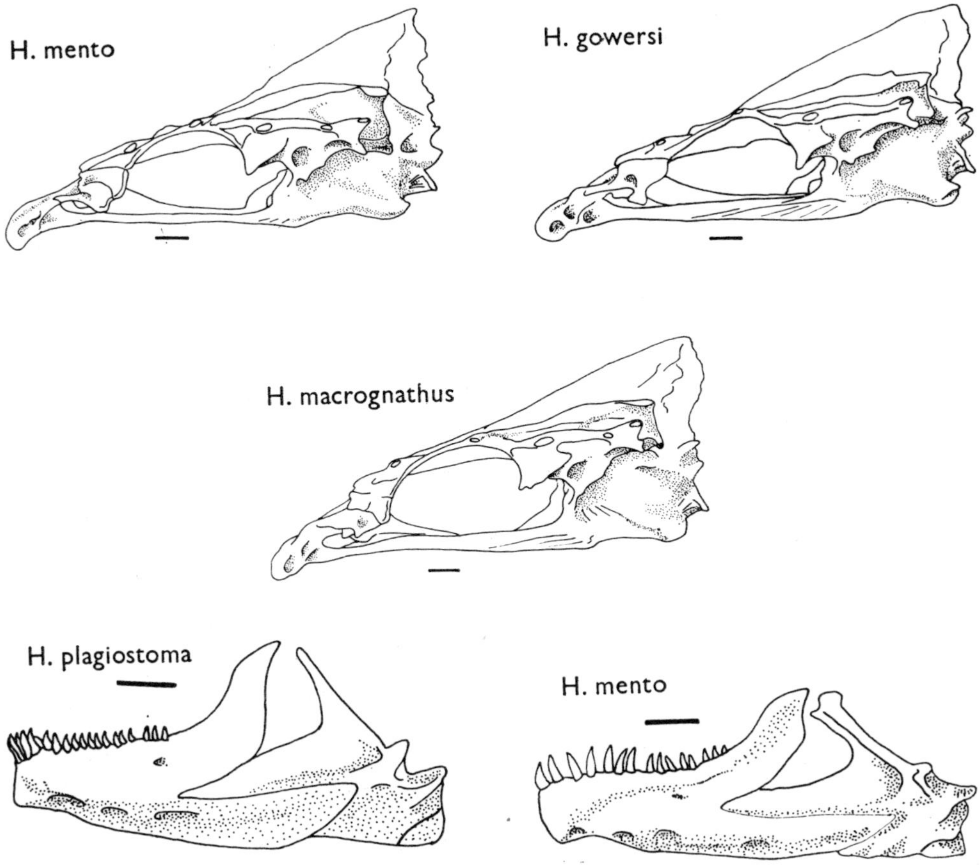

FIG. 57. Neurocranial and lower jaw form in piscivorous predators. The neurocrania are from members of the *H. prognathus* lineage, the lower jaws from members of the *H. serranus* (left) and *H. prognathus* (right) lineages. (Scale = 3 mm.)

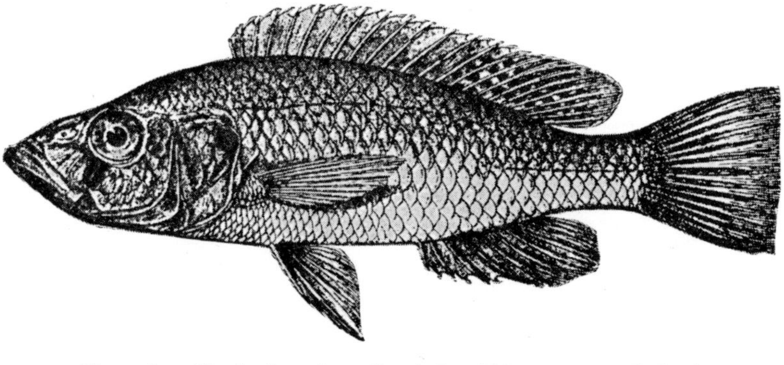

FIG. 58. *Haplochromis pellegrini*. (About natural size.)

specialized skull form than that occurring in the other two predator lineages (see Text-figs 51, 57, 68(4) and 69). The preotic length of the neurocranium is further elongate (65–68 per cent of neurocranial length, *cf* 60–65 per cent for the other lineages), the brain-case is shallower (28–32 per cent neurocranial length, *cf* 34–40 per cent), and the preorbital profile slopes at a much smaller angle (*cf* Text-figs 68(3) and 68(4) ; also Text-figs 47 and 51). Within the lineage a gradual intensification in this trend can be detected.

Two species (*H. bayoni* and *H. dentex* [Text-fig. 63] ; see p. 92 below) in the '*prognathus*' lineage can be grouped together on the basis of their having a skull in which the ethmo-vomerine region is noticeably decurved, although the rest of the skull retains the form typical for this lineage (Text-fig. 62).

Apart from these two species, all the others can, on the basis of increasingly specialized skull form, be collected into four subgroups. The phyletic interrelationships of the subgroups are but vaguely discernible. Indeed these categories show in cameo the difficulties involved in studying the phylogeny of the whole flock. Essentially, the problem is whether these intralineage groups are true, hierarchically evolved, sister groups (showing increasing apomorphy), or whether they represent gradal assemblages of polyphyletic ancestry (the ancestral species of each having, of course, reached the '*prognathus*' level of apomorphy). For the moment, this problem seems insoluble.

Species in each subgroup show some variation in gross morphology, although most are elongate, relatively slender fishes (the adult size range 140–200 mm), with a large mouth in which the inner and outer teeth are strong and unicuspid (even in small fishes, although some bicuspid outer and tricuspid inner teeth are found in the smallest fishes where these are known).

The first subgroup ('*paraguiarti*') has a neurocranial shape nearest that characterizing the '*serranus*' lineage (Text-figs 51 and 68(4) ; see p. 84). Of its seven constituent species, four, *H. paraguiarti* (Text-fig. 53), *H. gilberti* (Text-fig. 55), *H. pseudopellegrini* (Text-fig. 54) and *H. altigenis* are fairly similar in appearance, having a slightly oblique mouth and moderately slender body (Greenwood, 1967 ; Greenwood & Gee, 1969). Other species depart from this morphotype. *Haplochromis artaxerxes* has greatly elongate pectoral fins and a near horizontal mouth (Greenwood, 1962), *H. xenostoma* (Text-fig. 56) has a markedly oblique mouth (40–45 degrees with the horizontal) and a rather deeper body (Greenwood, 1967), while *H. flavipinnis* (Text-fig. 52B) also has an oblique mouth but with a deeply concave dorsal head profile (convex or straight in the other species ; Greenwood, 1962).

Neurocranial form in members of the '*paraguiarti*' subgroup is shown in Text-fig. 68(4).

Morphologically, species of the second subgroup ('*prognathus*', containing five species) are somewhat less uniform than those of the '*paraguiarti*' subgroup. *Haplochromis prognathus* (Text-fig. 13 ; Greenwood, 1967) and *H. bartoni* (Greenwood, 1962) are closely similar species, but *H. mandibularis* (Greenwood, 1962) has a narrower head, deeper cheek and more prognathous lower jaw, giving it a distinctive appearance seen again in a species of the following subgroup. *Haplochromis pellegrini* (Text-fig. 58 ; Greenwood, 1962) is of interest because of its small adult

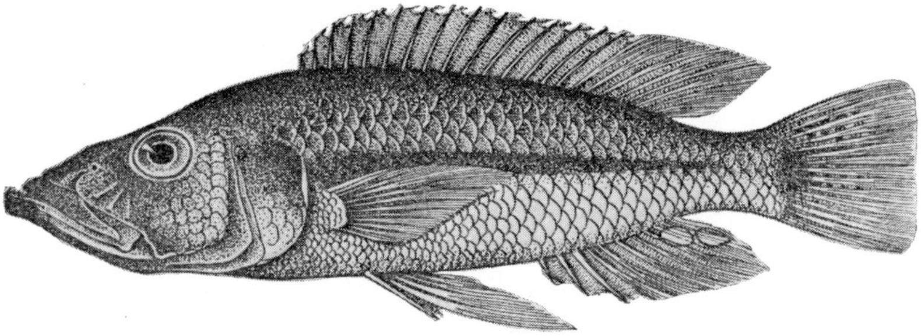

FIG. 59. *Haplochromis macrognathus*. (About half natural size.)

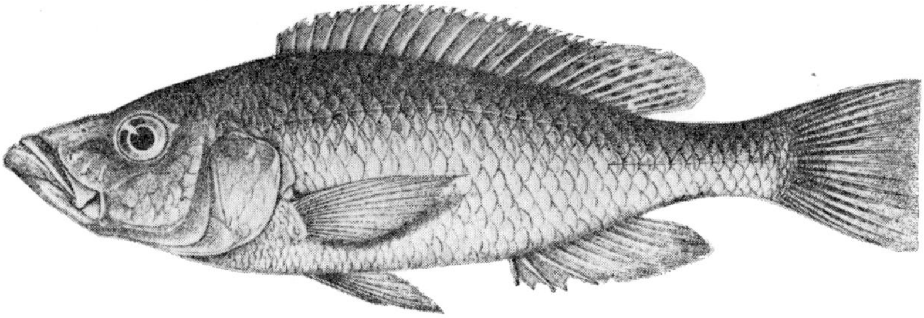

FIG. 60. *Haplochromis longirostris*. (About half natural size.)

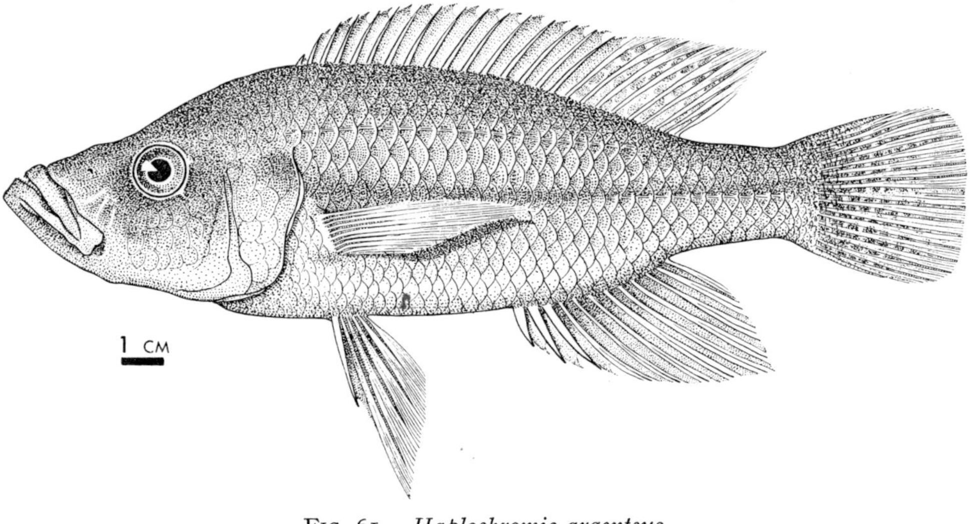

FIG. 61. *Haplochromis argenteus*.

size (71–105 mm). In its gross morphology *H. pellegrini* is rather like *H. paraguiarti* (of that nominal subgroup) but its neurocranial shape is virtually identical with that of *H. prognathus*; in other words, of a more specialized kind. The fifth member, *H. percoides* (Text-fig. 52A; Greenwood, 1962) also has small adults (70–95 mm), but its body shape and head form are quite unlike those of *H. pellegrini*; in these characters *H. percoides* closely approaches *H. flavipinnis* (of the '*paraguiarti*' subgroup), a species with which I previously considered it to be closely related (Greenwood, 1967). Like *H. flavipinnis*, *H. percoides* has a distinctive and unique colour pattern (*cf* Text-figs 52A and B).

Neurocranial form in members of the '*prognathus*' subgroup is shown in Text-figs 51 and 69(1).

The five species comprising the '*estor*' subgroup are perhaps the most striking and obviously predatory *Haplochromis* in the '*prognathus*' lineage. All have what may be considered a 'full' development of the skull type characterizing the lineage (Text-figs 57 and 69(2)), and all except *H. longirostris* (maximum adult size 145 mm) are among the larger piscivorous *Haplochromis* species in Lake Victoria.

Three species, *H. gowersi*, *H. mento* (Text-fig. 15) and *H. estor* (Text-fig. 14) are rather similar in appearance (Greenwood, 1962). A fourth, *H. longirostris* (Text-fig. 60; Greenwood, 1962), does not depart greatly from these species except for its rather terete habitus; skull shape in *H. longirostris* is, however, less extreme than in other members of the subgroup. *Haplochromis macrognathus* (Text-figs 57 and 59; Greenwood, 1962) is an extreme development of the subgroup morphotype (and superficially resembles *H. mandibularis* of the '*prognathus*' subgroup [see p. 89 above] but the head is narrower and the lower jaw is longer and more prognathous).

Neurocranial shape for the '*estor*' subgroup is seen in Text-figs 57 and 69(2).

The fourth subgroup of the '*prognathus*' lineage comprises a single species, *H. argenteus* (Text-fig. 61; Greenwood, 1967). This species, superficially, resembles *H. longirostris* (see above), but has a neurocranial form that is more anteriorly protracted than in any species of the '*estor*' subgroup (Text-figs 62 and 69(3)). This elongation, unlike that in the '*estor*' subgroup skull, involves relative protraction of the ethmo-vomerine region as well as the region of the skull between the prefrontals and the prootic.

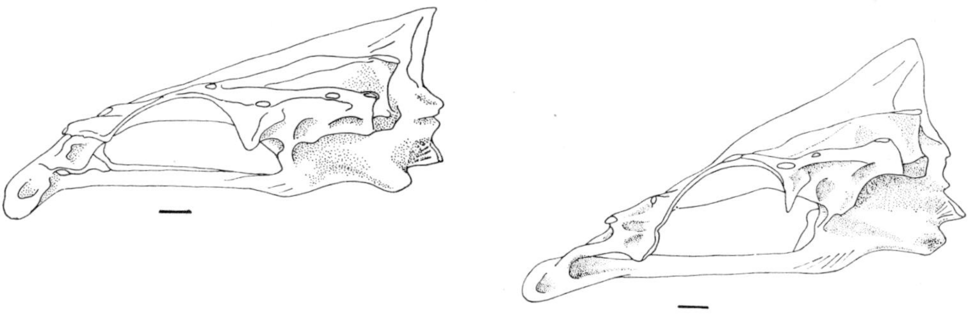

FIG. 62. Neurocranial form in *H. dentex* (left) and *H. argenteus*, two piscivorous predators of the *H. prognathus* lineage. (Scale = 3 mm.)

The relationship of *H. argenteus* to the '*estor*' species is undoubted, and its placement as a separate subgroup is only justified because of the unusual involvement of the ethmo-vomerine region in skull elongation.

Finally in this lineage there is the bispecific subgroup of *H. dentex* and *H. bayoni* (see p. 89 above), characterized by the fairly sharp decurvature of the ethmo-vomerine skull region (Text-figs 62 and 69(3)). This deviation from the typical skull form of the '*prognathus*' lineage is reflected in the almost horizontal alignment of the mouth in both species compared with the variously oblique jaws in other members (*cf* Text-figs 63 and 59–61). Besides this shared similarity in jaw alignment, both species are unusual in having relatively few (modal number 36–40) and larger outer jaw teeth. Judging from their overall neurocranial morphology, *H. dentex* and *H. bayoni* appear to be linked with the '*estor*' subgroup (see p. 91 above).

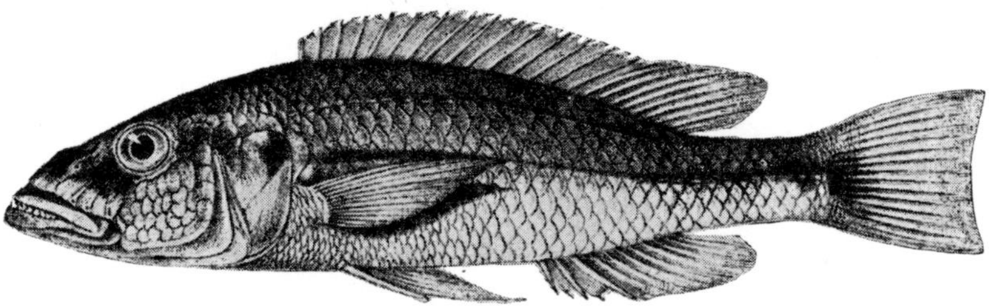

FIG. 63. *Haplochromis dentex.* (About half natural size.)

Four species, *H. dichrourous*, *H. apogonoides*, *H. orthostoma* and *H. parorthostoma* (Greenwood, 1967), have not so far been mentioned in this analysis, mainly because there is little information available on their cranial anatomy.

Haplochromis dichrourous is known to be piscivorous, but nothing is known about the diet of the three other species. The strong unicuspid teeth and large gape of *H. apogonoides* suggest that it may be a piscivore, as do the near vertical orientation of the jaws and the large gape in *H. orthostoma* and *H. parorthostoma* (Text-fig. 64).

Skull and jaw morphology in *H. apogonoides* is typically that of a generalized *H. macrops*-like species, although the stout, strongly recurved and unicuspid teeth resemble those of *H. sauvagei* (see pp. 69–70; also Greenwood, 1967). Since the skull form of *H. apogonoides* is not like that of the '*sauvagei*' group species (p. 70), and because its teeth could be interpreted as stouter versions of the kind found in *H. squamulatus*, the relationships of the species are probably with the latter group (see p. 85 above) and not, as I suggested previously, with the '*sauvagei*' group (Greenwood, 1967).

From the little evidence available (radiographs, partial dissection and overall morphology) *Haplochromis dichrourous* could well be a member of the '*prognathus*' lineage, probably of the '*paraguiarti*' subgroup (p. 89 above).

Radiographs of *H. orthostoma* and *H. parorthostoma* indicate a neurocranium of the *H. serranus*-type. Tentatively the species can be considered a derived offshoot of the '*serranus*' lineage (see p. 84 above). Certainly each is more closely related

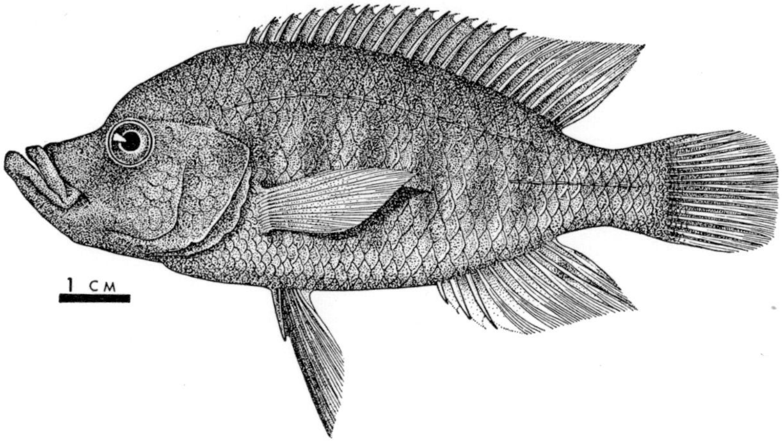

FIG. 64. *Haplochromis parorthostoma.*

to the other than to any other extant species. *Haplochromis parorthostoma* is known only from Lake Victoria, while its sister species, *H. orthostoma*, is confined to the Lake Kioga system.

To summarize this attempted phylogenetical survey of the piscivorous predators (Text-figs 65–70) we may note that, like other trophic groups, it is composed of more than one lineage. However, unlike the other groups, there seems to have been a greater phyletic evolution within the main piscivore lineage. One lineage (the *H. guiarti–H. squamulatus* line) may stem directly from a generalized insectivore of the '*pallidus*' type (see pp. 58–59). The other main lineage comprises two sister groups (the '*serranus*' and '*prognathus*' lineages) and seems to stem from an *H. saxicola*-like ancestor. The '*prognathus*' lineage is divisible into a number of subgroups probably of phyletic origin. No piscivorous lineage can be traced with certainty to an ancestor with *H. empodisma*-like affinities (see p. 63 above), although the possibility of the '*serranus*' group having such affinities cannot be ruled out (Text-fig. 70).

Conclusion

To close this discussion on the possible phyletic history of the Lake Victoria species flock, I would return briefly to the question of its mono- or oligophyletic origin, and also touch on the question of evolution above the species level in this and other species flocks.

In most trophic radiations (the paedophages [p. 75], lepidophages [p. 69], and possibly the piscivores [p. 80], excepted) the various member species seem referable to one of three lineages. The basic morphotypes of these lineages are represented today by the species *H. pallidus* (or *H. macrops*), *H. empodisma* and *H. saxicola*. All are insectivores, and the two latter species (with their close relatives) each represent a stage in the evolution of syncranial architecture from that of the presumed basic *H. bloyeti* type (see pp. 58–59).

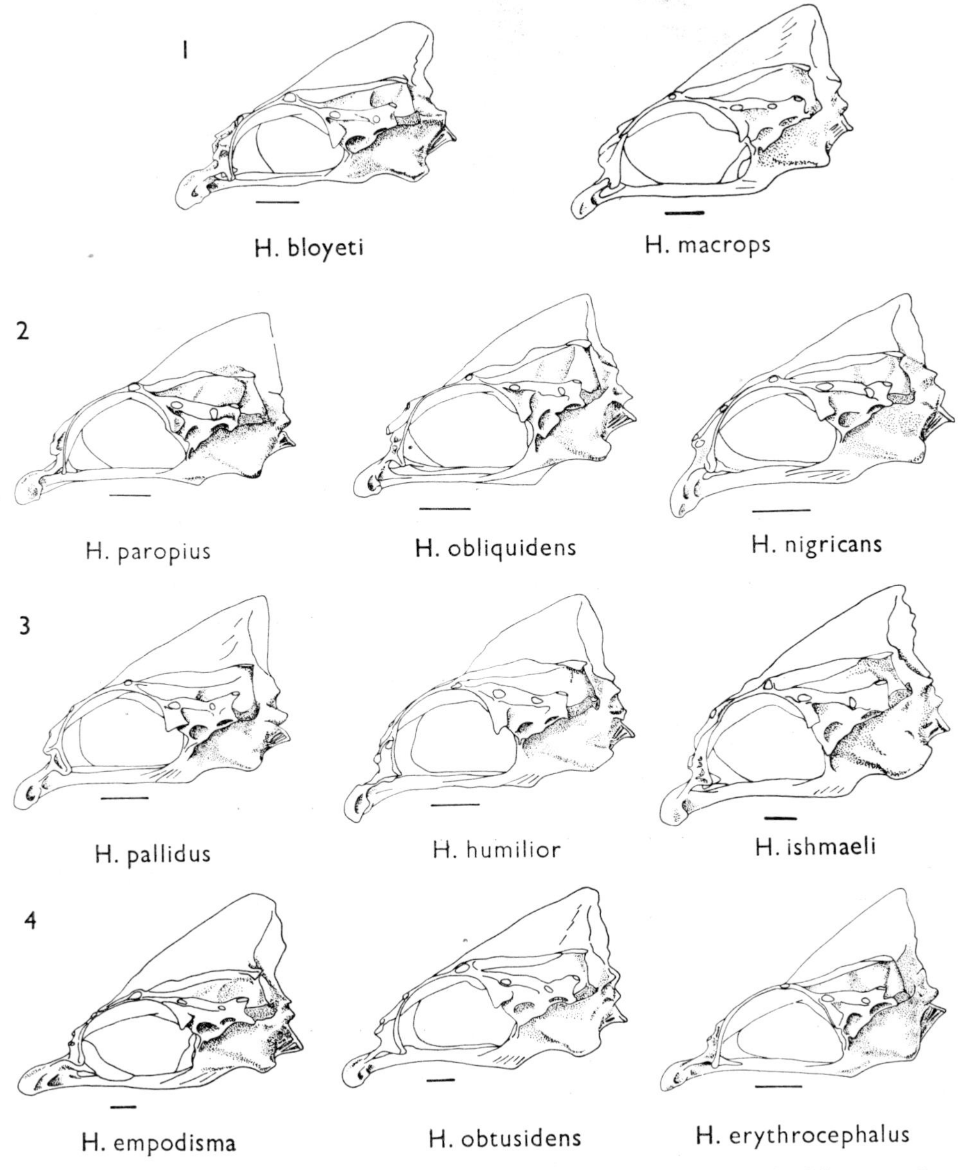

FIGS. 65–69. Neurocranial form in representative members of the principal lineages shown in the phyletic diagram (Text-fig. 70). (In all, scale = 3 mm.)

In this figure : 1. *H. bloyeti* and a member of the *H. macrops* line.
2. The *H. phytophagus–H. obliquidens* lineage.
3. The *H. pallidus–H. pharyngomylus* lineage.
4. The *H. erythrocephalus–H. obtusidens* lineage.

Species with syncrania of the *H. empodisma* and *H. saxicola* types are not found in the present-day east African river systems, whereas the *H. pallidus* type is well represented. In the absence of early Pleistocene *Haplochromis* fossils it is impossible to tell whether this situation also obtained in the rivers of that period, or whether the '*empodisma*' and '*saxicola*' skull types were then represented among fluviatile species.

Syncranial differences between *H. bloyeti*, *H. pallidus* and *H. empodisma* are, in fact, very slight (see Text-figs 31, 32 and 65). Conceivably, one of the earliest

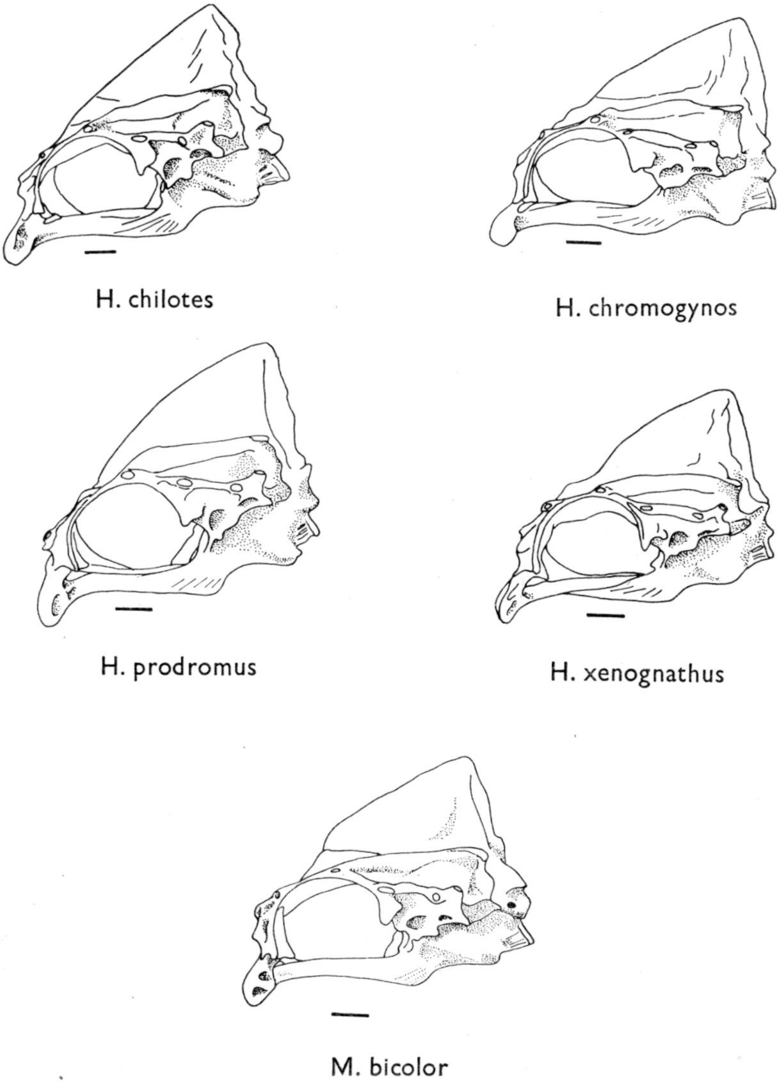

FIG. 66. The *H. chilotes–Macropleurodus bicolor* lineage.

dichotomies from an *H. bloyeti*-like stem species within the developing lake could have given rise to the *H. pallidus* and *H. empodisma*-like types. From this point a very slight differentiation of the *H. pallidus* skull type would give rise to the *H. macrops* type (see p. 59 and *cf* Text-figs 65(1) and (3)), and a somewhat greater change (but still only one of differential growth ; see p. 63) would produce, from the *H. empodisma*-like stem, a syncranium of the *H. saxicola* type (*cf* Text-figs 65(4) and 68(2)).

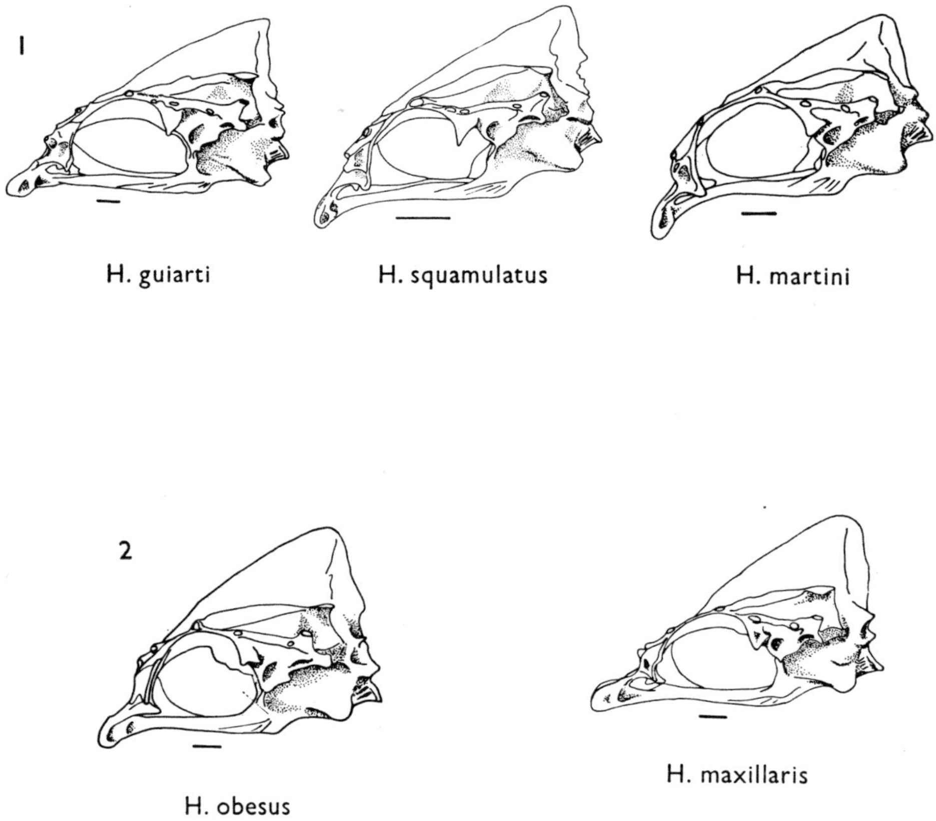

FIG. 67. 1. The *H. cinereus*–*H. squamulatus* lineage.
2. The *H. cronus*–*H. melanopterus* lineage.

Although a complete phyletic analysis of the flock cannot yet be made, it seems unlikely that additional material will alter substantially the scheme given in Text-fig. 70. Species still to be described will increase the number of terminal points categorized as a 'species complex' (or increase the number of species in those already recognized). They are unlikely to provide new lineages.

Even this partial phyletic analysis of the flock shows that, despite the superficial impression of virtually complete morphological and adaptational intergradation amongst its species, a number of distinct groups are discernible. For example there

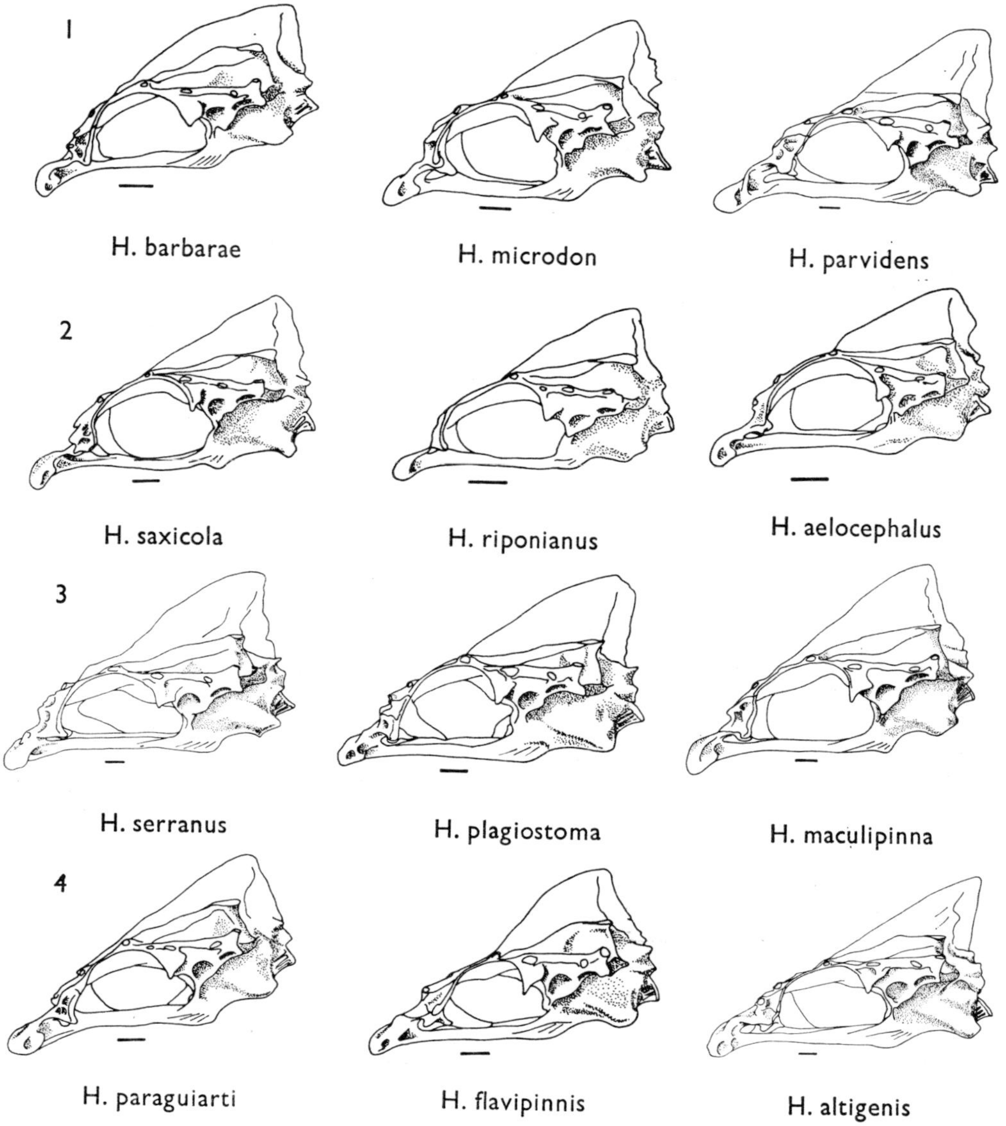

FIG. 68.
1. The *H. barbarae–H. parvidens* lineage.
2. The *H. riponianus–H. welcommei* lineage.
3. The *H. serranus* lineage.
4. The *H. prognathus* lineage ; the *H. paraguiarti* subgroup.

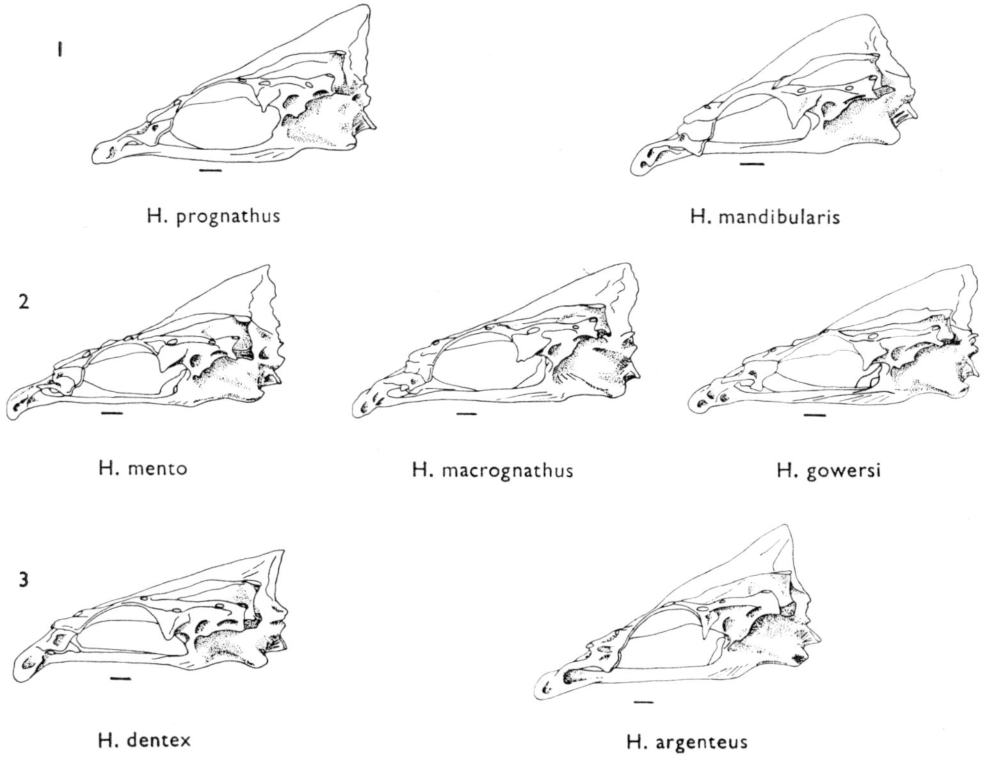

FIG. 69. The *H. prognathus* lineage continued.
1. The *H. prognathus* subgroup.
2. The *H. estor* subgroup.
3. The *H. dentex* subgroup (left) and the *H. argenteus* subgroup (right).

are the *H. pallidus–H. pharyngomylus*, *H. nuchisquamulatus–H. lividus*, *H. cinereus–H. squamulatus*, *H. sauvagei–H. chromogynos*, *H. cronus–H. melanopterus* and *H. barbarae–H. parvidens* groups, and the *H. serranus* and *H. prognathus* species complexes.

Each lineage is definable on the basis of certain shared morphological specializations and, on the whole, each is recognizable from its trophic peculiarities (see Text-fig. 7). The distinguishing characters would be sharpened if the least specialized members of the groups had been eliminated in the past (as may have happened in the history of *Hoplotilapia* and *Platytaeniodus*). In other words, we seem to have species aggregates which might be accorded higher rank.

That the origin and evolution of higher categories involve processes no different from those involved in speciation is now generally accepted. The real problem in this particular situation is to find some satisfactory definition of the genus, a definition that is not purely a pragmatic one like that which Mayr (1969) proposes. The size of the morphological gap (Mayr's criterion) separating two taxa does not necessarily indicate the degree of their phyletic divergence (for example, take *Macropleurodus bicolor*; see p. 71 and Text-fig. 75).

That a genus must be monophyletic is self-evident, but the points where 'generic' lines are to be drawn are by no means evident in the case of a species flock. For example, when discussing this problem with my colleagues, one suggested that a genus should comprise, at most, the two apomorph and the two plesiomorph sister species arising from three dichotomies. This definition could not be applied to a situation where there was multiple synchronous speciation from a common ancestral species isolated in several places (as may well have happened in the evolution of the Victoria species flocks; see pp. 108–114). In other words, at an early stage in the evolution of this flock each of several populations from the same ancestral species could have given rise to very differently specialized lineages. Polychotomous rather than dichotomous branching could be basic to the evolutionary history of this and other lacustrine cichlid species flocks. Similar branching patterns could also have occurred within a lineage, thereby giving rise to the clusters of very similar species at any one level of apomorphy or plesiomorphy.

Because for the moment I see no solution to this 'generic problem', the various lineages indicated in Text-fig. 70 are left within a single but almost certainly artificial genus.

A large question mark also hangs over the phyletic relationship of the Lake Victoria *Haplochromis* species with those of Lake Malawi. The generalized species in both lakes are alike in all anatomical features, and of greater import, both flocks have species with identical specializations. Although this would seem strongly to indicate close similarity in the genotypes of their ancestral species, it cannot be established if these taxa were conspecific. Further arguments against the ancestors being conspecific are the greater age of Lake Malawi (see p. 6), and the occurrence in Lake Malawi *Haplochromis* of certain colour patterns and types of squamation not seen in any Lake Victoria species (p. 20). On a truly phyletic basis, therefore, it is problably wrong to place any of the Lake Malawi species in the same genus (or genera) as those from Lake Victoria.

Both these 'generic' problems are under investigation, as are the inter- and intra-relationships of the Lake Tanganyika Cichlidae.

THE RELATIONSHIPS OF THE MONOTYPIC GENERA

The five monotypic genera, *Astatoreochromis alluaudi* (Greenwood, 1959a), *Hoplotilapia retrodens*, *Platytaeniodus degeni*, *Macropleurodus bicolor* and *Paralabidochromis victoriae* (see Greenwood, 1956a) are all clearly derived from *Haplochromis* ancestors. Except for *Macropleurodus bicolor* and *Astatoreochromis alluaudi*, however, it has so far proved impossible to suggest with which extant *Haplochromis* species these genera show the closest affinities.

The relationship of *Macropleurodus bicolor* (Text-fig. 3) with the *H. sauvagei* group of mollusc eaters is discussed on p. 71.

Arguments are presented elsewhere (Greenwood, 1954, 1959a; also see p. 18 above) that *Astatoreochromis alluaudi* originated from a stem distinct from that of all other Lake Victoria haplochromine species. Apparently the nearest living

FIG. 70. A tentative phylogenetic arrangement of the Lake Victoria *Haplochromis* species flock based on the assumed synapomorphy of various character states (particularly those of the neurocranium and the dentition); see text pp. 57–99. Although each lineage (except the *H. macrops* complex) is recognized on the basis of specialized characters shared by its constituent species, no shared specializations have yet been found that will enable one to link, dichotomously, the different lineages to a common ancestral species. That is to say, the basal members of a particular lineage show only specializations interrelating them with members of that particular lineage; in other respects they are at a generalized (plesiomorph) level of organization. Each could have been derived (and derived synchronously) from an *H. bloyeti*-like ancestor (see text, pp. 57–58).

It has also proved impossible to interrelate constituent species of the *H. macrops*, *H. serranus*, *H. tridens* and *H. prognathus* complexes. In the three latter groups all species show an equally derived level of organization, while in the *H. macrops* complex the species are so generalized that synapomorphies are not detectable (and thus the association may not constitute a truly phyletic assemblage).

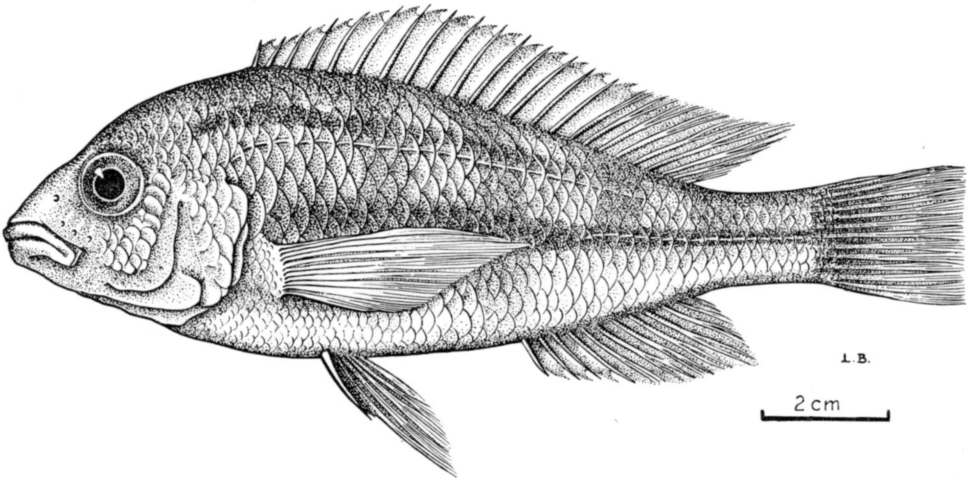

FIG. 71. *Platytaeniodus degeni.*

relative of *A. alluaudi* is a small, mollusc-crushing species *Haplochromis vanderhorsti* from the Malagarasi river system of Tanzania (Greenwood, 1954); it differs from *Astatoreochromis alluaudi* mainly in having the typical *Haplochromis* number of anal fin spines (3), rather than the elevated number found in *Astatoreochromis*, *viz* 4–6.

As there is but one specimen of *Paralabidochromis victoriae* available for study (Greenwood, 1956a) only superficial characters can be used to assess its relationships. The peculiar, procumbent, elongate and unicuspid teeth of this species, coupled with its narrow dental arcade, seem to imply relationship with *Haplochromis chilotes*, a specialized and insectivorous member of the *H. sauvagei* lineage (see p. 61 above). This conclusion cannot, however, be tested until osteological material becomes available for comparison.

For *Hoplotilapia retrodens* and *Platytaeniodus degeni* (Text-figs 71 and 72), despite adequate study material, it is impossible to suggest any living species or species

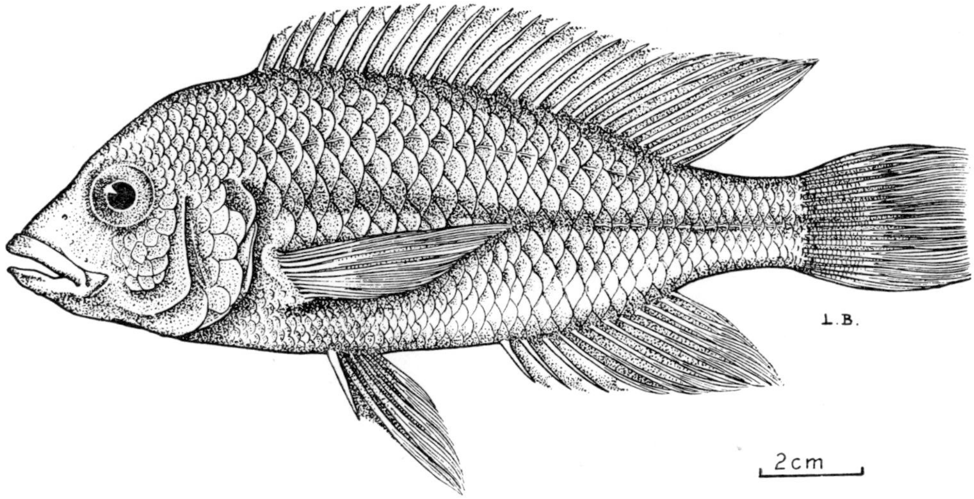

FIG. 72. *Hoplotilapia retrodens.*

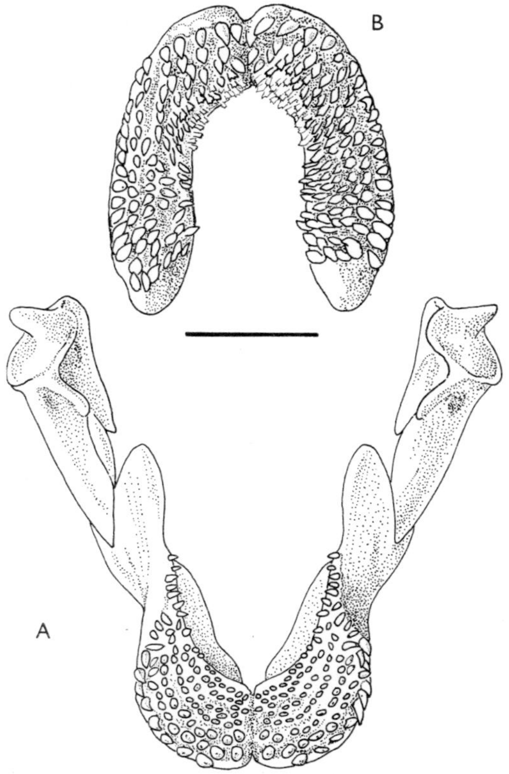

FIG. 73. *Platytaeniodus degeni.* A: Lower jaw in occlusal view. B: Dental pattern of the premaxilla; in larger individuals the posterior part of tooth band on each side is expanded medially. (Scale = 3 mm.)

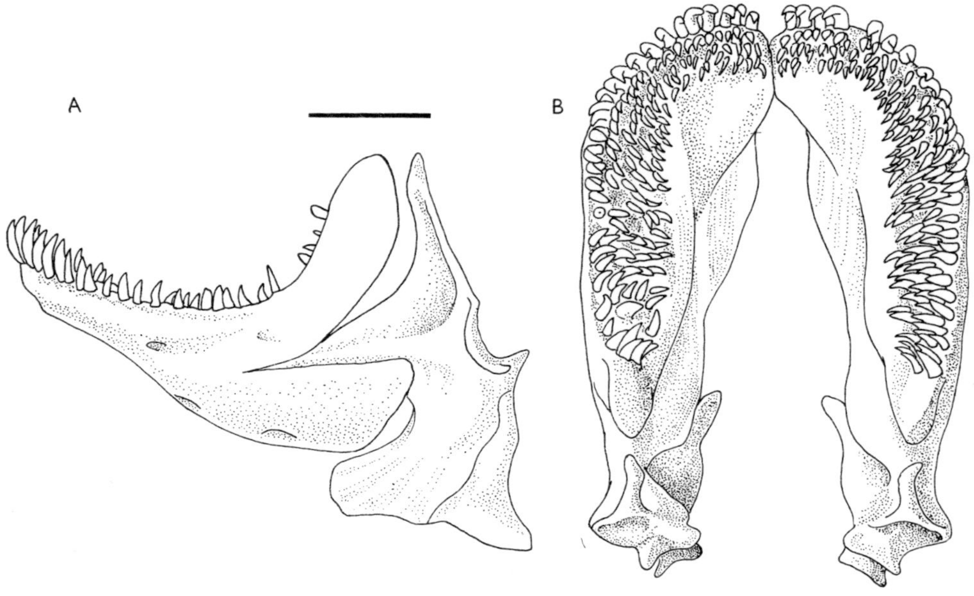

FIG. 74. *Hoplotilapia retrodens.* Lower jaw in lateral (A) and occlusal (B) views. (Scale = 3 mm.)

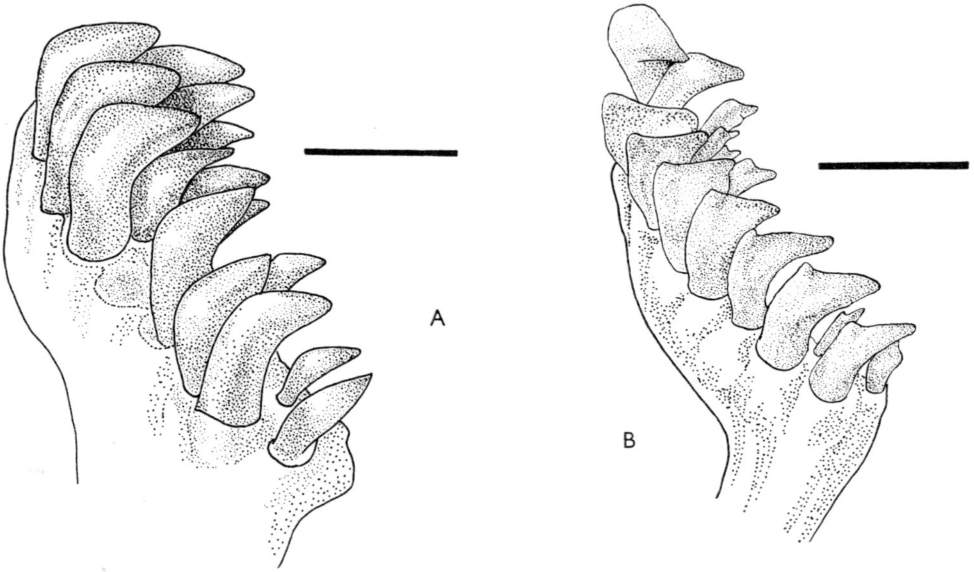

FIG. 75. *Macropleurodus bicolor.* Left premaxilla, occlusal view seen from above and medially. A : From a large fish (a male, 115 mm standard length), in which the minor cusp is absent from all teeth. B : A smaller individual (a female, 88 mm S.L.) showing the prominent minor cusp. (Scale = 1 mm.)

group as the nearest relative. An earlier view (Greenwood, 1956a) that both taxa are related to the *H. crassilabris* species complex of Lake Victoria can no longer be substantiated on either dental or neurocranial similarities. *Hoplotilapia* and *Platytaeniodus* have, as their diagnostic features, a great widening of the tooth bands in both the dentary and the premaxilla (Text-figs 73 and 74), combined with a tendency for teeth of the inner rows to be as large as those in the outer row. Although premaxillary shape and tooth arrangement is basically similar in both genera (particularly in small individuals ; see Greenwood, 1956a) the shape of the dentary and its tooth bands are markedly different (*cf* Text-figs 73 and 74). In *H. retrodens* the dentary is flattened and the broad bands of teeth extend posteriorly onto the ascending coronoid process (a feature not encountered in any other Lake Victoria haplochromine). In *P. degeni* the bone is deep and stout anteriorly, and the multiseriate teeth are virtually confined to its anterior third. Tooth shape also differs intergenerically ; in *P. degeni* the crowns of the unicuspid teeth are flattened and resemble those of a typical bicuspid, whereas in *H. retrodens* the teeth, also unicuspid in adults, are conical and recurved.

Neurocranial form is identical in *Hoplotilapia* and *Platytaeniodus* (Text-fig. 76) and approaches the generalized *H. macrops* type. But no member of the *macrops–pallidus* lineage (see p. 58) shows even an incipient development of the dental features characterizing either genus. This absence of anatomically annectent species is, as will be recalled, an outstanding feature in this species flock where all other structural grades are completely bridged by intermediate forms.

Now that more is known about the anatomy and ecology of the whole flock, my original interpretation (Greenwood, 1956a) of the monotypic genera as being the results of quantum evolution (*sensu* Simpson, 1953) hardly seems tenable. There are numerous examples within the flock of species that, trophically and morphologically, occupy intermediate positions with respect to different adaptive (i.e. trophic) zones, and there are several species in the mollusc-eating grade to which *Hoplotilapia* and *Platytaeniodus* belong. In other words, these genera do not occupy a niche or adaptive grade unoccupied by a close relative, one of the features characterizing quantum evolution. Nor, with respect to the absence of morphological intermediates, is it clear why those linking *Platytaeniodus* and *Hoplotilapia* should have been any more ill-adapted (and therefore rapidly exterminated) than those annectent forms still surviving in other lineages.

The morphological isolation of at least these two monotypic genera remains a mystery (with undertones, almost, of Goldschmidtian 'macromutations' and 'hopeful monsters'). The generic status of *Macropleurodus* and *Paralabidochromis* is, I now think, questionable.

THE ANATOMICAL BASIS FOR THE ADAPTIVE RADIATION

It is chiefly in their cranial anatomy that one sees the results of the adaptive radiation undergone by the Lake Victoria *Haplochromis* species during the last three-quarter million years.

Some of the interspecific differences seem profound. Yet, on careful examination the different characteristics are found to be relatively simple variants of a basic

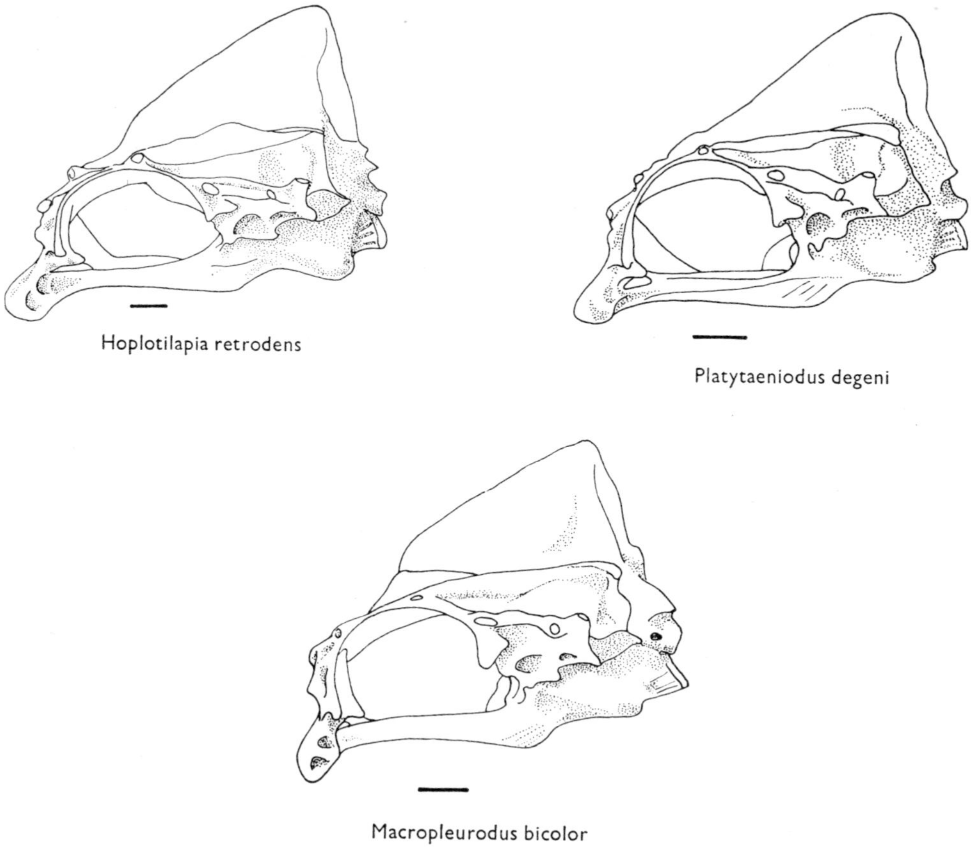

FIG. 76. Neurocranial shape in three endemic monotypic genera. (Scale = 3 mm.)

bauplan. The relative simplicity of these morphological changes, coupled with the basic anatomical level already reached by the ancestral lake cichlids, has, I believe, given the haplochromine fishes a great 'morpho-potential' with which to meet the environmental opportunities provided in a developing lake (Greenwood, 1973b).

Alexander (1967) first drew attention to certain peculiarities in the jaw mechanism of a South American cichlid, *Pterophyllum*. Subsequently I have been able to show that this specialized type of jaw mechanism occurs in all the Lake Victoria *Haplochromis* species I have dissected (i.e. the majority of species, including most monotypic genera).

These specializations concern functional interrelationships of the premaxillae and maxillae, the effect of which is to allow greater protrusion of the upper jaw, and firmer fixing of that jaw in the protruded position (see Text-fig. 77). As a result, the lower jaw can be closed firmly against the upper jaw whilst the latter is still protruded. The advantages of this system are discussed in detail by Alexander (1967). They are further exemplified by the exploitation of this system in some groups of mollusc eaters (the shell-wrenchers, p. 69), a type of feeding habit that could hardly evolve without the cichlid-type of protrusion mechanism. Its advantages are also seen variously exploited in other benthic feeders, and again in the great piscivorous predator radiation (including, particularly, the paedophagous species; p. 75 *et seq.*). Indeed, the functional flexibility of the cichlid type of jaw mechanism is probably one of the prime factors leading to the evolutionary success of these fishes.

With few exceptions (e.g. the monotypic genera [Text-figs 73–75] and some of the paedophages [Text-fig. 46, and p. 77]) there is relatively little variation in the shape of the jaw bones. In the molluscivores that crush snail shells orally, or wrench the body from its shell, the dentary (Text-figs 40 and 41) is noticeably shorter and

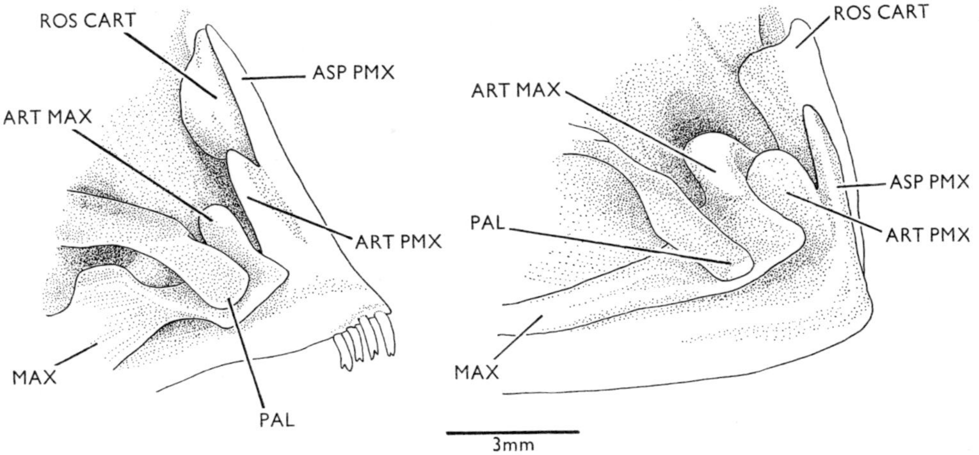

FIG. 77. View, from a dorsolateral position, of the right upper jaw in: *Perca fluviatilis* (Right) and: *Haplochromis brownae* (Left). The associated ligaments and other connective tissues, and the cartilaginous menisci between the various elements are omitted. The rostral cartilage, however, is shown *in situ*.

In *Perca*, the articular process of the maxilla (ART MAX) is in direct contact with the dorsal and dorsomedial aspect of the premaxillary articular process (ART PMX) through a cartilaginous meniscus, and also partly overlaps the posterolateral face of the same process. Additional contact between the articular process of the maxilla and the premaxilla is brought about through its contacting the rostral cartilage (ROS CART) underlying the ascending process of the premaxilla (ASP PMX). As a result of these contacts, protrusion of the premaxilla is at least partly effected through the rotation of the maxilla (see Alexander, 1967, and also Liem, 1970). In *Haplochromis* (all species examined, see p. 104) the articular process of the maxilla only contacts the premaxilla at one point, through a cartilaginous meniscus with the ventrolateral face of the premaxillary ascending process (i.e. anterior to the rostral cartilage). Unlike *Perca* there is neither contact with the articular process of the premaxilla (ART PMX) nor with the rostral cartilage.

Protrusion of the premaxilla thus seems to be effected mainly through the sinking of the lower jaw, to which the premaxilla is linked by the connective tissue of the lips. When the premaxilla is fully protruded, the articular process of the maxilla (ART MAX) does, however, contact the upper part of the premaxillary articular process. In this way it helps in holding the upper jaw in the protruded position even when the lower jaw is raised (see text p. 107).

the premaxillae somewhat stouter than in other species. In certain lineages (especially those adapted for preying on other fishes) the jaws may be relatively longer, as may the ascending processes of the premaxillae, but the overall morphology of these bones shows little change (*cf* Text-figs 31 and 57).

Liem (1970) has demonstrated the importance of a high 'coronoid' process on the dentary for increasing the speed at which both the jaws can be closed. In all the Lake Victoria *Haplochromis* species examined (and in the generalized *H. bloyeti*) the coronoid process could be classed as 'high'. It is relatively a little higher in certain species, particularly among the advanced piscivores and in both lineages of paedophages. In many of these species (and particularly among the piscivorous predators) the lower jaw is relatively longer and the upper jaw markedly protrusile, a combination of characters enhancing the grasping range and rate of the jaws, and increasing the volume of the buccopharyngeal cavity (Liem, *op. cit.*).

In general, however, it could be said that jaw form is relatively constant throughout the flock; what interspecific differences there are can, in most instances, be attributed to the results of allometric growth changes.

The high coronoid process of all *Haplochromis* species can be considered as one of the basic features contributing to the success of the *Haplochromis*-bauplan as raw material for an adaptive trophic radiation.

Basically, only two types of outer jaw teeth occur throughout the Lake Victoria species flock (see Text-fig. 4), namely, teeth with bi- or unicuspid crowns. The great variety of bicuspid crowns, including the extremes in shape and development of one cusp found in algal grazing species (Text-figs 4D–G), are simply the products of relative growth, as is the degree to which the tooth neck is curved inwards or even outwards (Text-figs 4, 33 and 46).

Except in their degree of curvature, unicuspid teeth show little variation in shape. Unicuspid teeth are usually taken as the mark of a piscivorous predator, and their adaptive significance in this group is self-evident. However, an analysis of tooth form in all species shows, with few exceptions, that the presence and relative abundance of unicuspid teeth is generally a correlate of the fish's size. In other words, individuals in all species have bicuspid teeth until a length of *ca* 80–90 mm is reached; above that length there is a size range (80–110 mm) in which the dentition is a mixture of bi- and unicuspids. This is followed by a wholely unicuspid dentition in larger individuals. Since the maximum size attained by fishes in the majority of non-piscivorous species is rarely greater than 100 mm, it follows that the dentition in these species is generally a bicuspid one.

The only exceptions I can find to this generalization are the algal grazing species (see pp. 63–66) where, as might be expected, the specialized bicuspid tooth (Text-figs 4D–G and 36) is retained even in the largest fishes (although individuals larger than 95 mm are rare). Also exceptional are the very few piscivorous species where unicuspid teeth predominate in fishes less than 80 mm long, irrespective of the maximum adult size attained (be it more or less than 100 mm).

Thus, generally speaking, in the evolution of piscivorous fishes an increase in adult size would have the double adaptational value of providing a unicuspid dentition and an increase in speed (since larger fishes swim faster than smaller ones). Clearly, some piscivores have succeeded in breaking away from this size-correlated change in dental form, as witnessed by the few exceptional species mentioned above, but this is unusual.

Inner row jaw teeth show less variation in shape than do those of the outer row. The main differentiation here is in the disposition of the teeth ; that is, a relative increase or decrease in the number of tooth rows and the area over which the widening of a tooth band is manifest. Obvious functional correlation with increased width of inner tooth bands is seen in the algal grazers (p. 65), some mollusc eaters (p. 71 ; Text-fig. 41) and in the scale-eating *H. welcommei* (p. 69 ; Text-fig. 39).

The pharyngeal bones and dentition show considerable variation in form (see Text-figs 5 and 30). Here again the changes involved are simple ones concerning, chiefly, an increase in the stoutness and robustness of the lower pharyngeal bone and a differentiation of its dentition. The upper pharyngeal bones show a correlated change, but usually the changes are less marked.

The principal pharyngeal dental changes involve either the replacement of compressed, bicuspid teeth by flat-crowned, stout teeth, or a coarsening or refining of the teeth without a change in their basic bicuspid form. The change from a laterally compressed pharyngeal tooth to a molariform one would seem to be greater than any seen in the oral dentition (see above). Furthermore it is not simply correlated with the fish's size, as is the change from bi- to unicuspid jaw teeth (although the degree of molarization in a pharyngeal dentition is often positively correlated with the fish's length). Strengthening of the pharyngeal bones is, in general, correlated with the change from a basic bicuspid dentition to one in which molariform teeth are present. Other variants in the pharyngeal mill pattern involve the shape of the dentigerous area on the lower bone (always basically triangular, however), and the spacing of the pharyngeal teeth on it.

Changes in neurocranial form have been described and figured in the preceding chapter (pp. 93–99 ; Text-figs 65–69). Their importance in trophic adaptation is seen especially in the ethmo-vomerine region, the base on which the upper jaw moves and is supported. Thus it is not surprising that changes in the shape and relative proportions in this region of the skull are so obvious in this *Haplochromis* flock. Through alteration to the angle at which the ethmo-vomerine region slopes, changes are effected in the striking angle of the upper jaw during protrusion ; relative changes in ethmo-vomerine length influence the degree to which the premaxilla can be protruded whilst still retaining a firm foundation, and it is this region of the skull that plays an important role in allowing the lower jaw to bite hard against a still protruded upper jaw (see p. 105 above). The involvement of all these different factors can be seen, individually or in combination, in the skull forms of various *Haplochromis* species.

In the '*sauvagei*' radiation of mollusc eaters (pp. 69–72) we clearly see all three factors involved. The steep ethmo-vomerine slope (Text-figs 40 and 66) induces a downward as well as the usual forward protrusion of the upper jaw. This region of the skull is relatively elongate so that the ascending processes of the premaxillae are strongly supported at all stages of protrusion. Finally, the lower jaw can shut quickly and firmly against the upper jaw whilst it is still protruded, thus allowing the fish to hold the snail shell firmly whilst it either wrenches the shell away from the body or crushes it between the upper and lower jaws.

The morphological changes that have allowed the development of this particular feeding habit represent simple alterations in the growth pattern of a generalized (i.e. *H. macrops*-like) syncranial type, and are especially obvious in the neurocranium (*cf* Text-fig. 65(1) with Text-fig. 66).

Relative growth changes in parts of the neurocranium are particularly well illustrated in the '*prognathus*' lineage of the piscivore radiation (see pp. 85–92). Here the changes involve a relative lengthening of the entire preotic skull region,* a decrease in the height of the neurocranium and, as a consequence, a marked decrease in the angle at which the preorbital face of the skull slopes downwards (*cf* Text-fig. 68(4) with Text-fig. 69(1–3)). In turn, this gentler slope of the path over which the premaxillary processes slide results in a different orientation of the toothed premaxillary arms as compared to the situation in a generalized skull. The ascending processes and the toothed arms of the premaxillae meet at almost a right angle in all species ; consequently in fishes with the skull shape under consideration, the dentigerous arms, relative to those in a generalized syncranium, lie at a steeper angle to the horizontal (because the ascending processes lie more nearly horizontally on the flatter ethmo-vomerine region). Correlated changes in the relative length of the lower jaw (inevitable if the system is to remain functionally integrated) result in a larger gape than is possible in fishes with a generalized neurocranial shape. Furthermore, the gape is directed forward and horizontally (advantageous in a hunter) and not obliquely downwards as in a generalized syncranium (a condition, nevertheless, suited to benthic feeding habits). Possibly the chief gain of the former mouth type to a predator is in being able to take prey at the same level as the fish's line of forward progress, rather than by having to align its body away from the horizontal in order to bring the gape into that plane.

The cichlid type of premaxillary–maxillary relationships (see Text-fig. 77 ; also Alexander, 1967), enabling the fish to brace the protruded premaxillae and to raise its lower jaw with the upper jaw fully protruded, seems to be of particular significance in the piscivore radiation. First, it allows a slightly greater protrusion of the upper jaw than does the basic percoid arrangement (Alexander, 1967) and thus an increase in the volume of the orobranchial cavity. Second, it allows the prey to be firmly gripped between the jaws without a marked decrease in the orobranchial volume (as would happen if the upper jaw has to be retracted before the lower jaw closed on it). This latter characteristic is of particular importance because cichlids do not swallow their prey whole. Instead, the body of the prey is gradually rasped away by the pharyngeal mill posteriorly while the rest is held by the jaws and orobranchial cavity.

Theoretically, this combination of pharyngeal mastication, firm jaw grip and increased expansion of the orobranchial cavity would permit the capture and ingestion of larger prey than would be possible if ingestion was simply by swallowing the prey

* i.e. The skull from the anterior tip of the vomer to the anterior margin of the prootic, and thus including the frontals and the parasphenoid.

whole. Unfortunately this idea cannot be tested because no figures are available for the relative differences in prey sizes of cichlids and species bolting their prey without pharyngeal mastication. However, subjective appraisal certainly suggests that the cichlids do eat relatively much larger prey than does, say, the catfish *Clarias mossambicus*, although the latter will probably consume more individuals per unit feeding time (personal observations).

To return to the modifications in neurocranial form seen in piscivorous *Haplochromis*. As was the case in the molluscivorous '*sauvagei*' radiation (see above), the alterations in relative skull proportions are a reflection of differential growth changes. Correlated modifications in jaw size and in proportions of the suspensorium are also due to changes in relative growth. Growth related changes are also seen in the increased streamlining of body form and head shape (Text-fig. 3) and, more directly, the increased body size. Many changes are, of course, allometrically correlated with this increased overall size and so too is the shift from a bicuspid to a unicuspid oral dentition (see p. 106 above).

Apparently without exception then, the underlying morphological principal involved in the evolution of this trophically complex and morphologically varied species flock is one of differential growth in various elements of the syncranium. Included in this generalization are, of course, the muscular as well as the bony components. Less work has been done on the cranial musculature than on the skeleton of these fishes. The observations I have made suggest that throughout the flock the chief myological differences are also proportional ones, coupled with slight differences in the insertion points of certain muscles, especially the adductor mandibulae series.

As far as I can tell from published accounts (and from limited personal experience) the same generalizations can be made about the osteology and myology of the Cichlidae from Lakes Malawi and Tanganyika. Certainly they apply to those of Lakes Edward, George and Albert. In Lakes Malawi and Tanganyika some species (generally referred to genera other than *Haplochromis*) do exhibit a more profound change in the form of the outer, less frequently the inner, jaw teeth (compare Text-fig. 4 with text-figs 28 and 29 in Fryer & Iles, 1972). Nevertheless, the derivation of these teeth from a basic bi-, tri- or unicuspid type is evident and of the same nature as that seen in the Lake Victoria species.

The importance of changed habits and behaviour in the origin of new adaptive trends is rightly emphasized by Fryer & Iles (1972 : 207 and 486). But their statements that these 'produce . . . changes in function which themselves lead to changes in structure . . .', and again '. . . differences in habits and behaviour can precede and initiate morphological differences . . .' are, to say the least, equivocal. Possibly they had in mind the example provided by *Astatoreochromis alluaudi* where differences in diet have pronounced effects on the degree to which certain specialized characters are developed (Greenwood, 1959a, 1965c). Fishes deprived of their usual diet of gastropods, or those feeding on snails with thin shells, have less well-developed pharyngeal bones, fewer molariform pharyngeal teeth, and a much reduced apophysis for the upper pharyngeal bones on the skull base. Yet, despite their reduced state, all three specialized characters are present and are recognizable as such.

Many *Haplochromis* species in Lake Victoria are facultative and opportunistic feeders (personal observations). During the early stages of the flock's evolution this ability could have had important survival value in isolated populations. Such behaviour might include incipient trophic specialization in an otherwise generalized feeder and thereby lay the foundation for the selection of specialized anatomical features. But apart from *Astatoreochromis alluaudi* nowhere amongst the Lake Victoria *Haplochromis* have I found evidence of unequivocal phenotypic modification in response to an environmental influence.

The evidence from *Astatoreochromis alluaudi*, however, strongly suggests that the supposedly 'reduced' condition found in 'atypical' populations may represent the full extent of genetical control over the development of these specialized characters. The condition in other populations (the 'typical' ones) would then be an environmentally influenced hypertrophy of the genetically controlled condition. The reason for reaching this conclusion is the appearance, in a few generations, of the 'reduced' characters in an aquarium population descended from parents showing full hypertrophy of these features (Greenwood, 1965c).

A contradictory situation is found in *H. ishmaeli* of Lake Victoria and a very closely related species, *H. mylodon*, of Lakes Edward and George (Greenwood, 1973a). In Lake Victoria *H. ishmaeli* feeds mainly on the hard-shelled gastropod *Melanoides tuberculata* (also the food of *Astatoreochromis alluaudi*). In Lakes Edward and George, *H. mylodon* feeds on both insects and molluscs, the latter thin shelled and usually of genera other than *Melanoides* (Greenwood, 1973a). Yet, the pharyngeal bones and skull apophysis in the two species show the same degree of hypertrophy. Here it would seem that the major factor controlling development of the specialized characters is a genetical one.

Even though we are far from an understanding of the epigenetical mechanisms influencing ontogeny, the results of these processes, as argued above, are interpretable in terms of changes in relative growth. Taken in conjunction with the basic syncranial and dental characters of a generalized haplochromine cichlid, this relatively simple mechanism could explain, in large part, the evolutionary success of these fishes in the African lakes (see also p. 111). It could also explain the rapidity with which the Cichlidae have been able to produce the adaptive modifications underlying their ecological dominance in the fish faunas of these lakes.

Another factor that may have contributed to rapid adaptive radiation in the changing environments of a developing lake is the nature of the dental modifications, both oral and pharyngeal.

Judging from the 'intermediate' morphology still preserved in species of some lineages (especially the mollusc crushers, the periphyton grazers and the mollusc shellers) the change from a generalized tooth form could have immediate or almost immediate selective value. It would not, therefore, have to undergo any (or much) modification before reaching a preadaptive stage with respect to its new function. In other words, selection forces acting on it would be essentially postadaptive rather than the usual succession of pre- and postadaptive ones (see Bock, 1959).

The outstanding explosive radiations of cichlid fishes (particularly haplochromine species) in developing lakes may well be consequent upon there being an abundance of selection forces for new structures (as compared with the pre-existing and stabilized river situation) and the likelihood that slight morphological changes (in, of course, the right synanatomical context) are sufficiently preadaptive for these forces to become immediately effective.

In essence, the success of the haplochromine cichlids seems attributable to their fluviatile ancestors having reached a level of general specialization which could be further developed and differentiated by simple ontogenetic reorganization. Anatomically the other fishes were either too specialized (e.g. the catfishes, the Cyprinidae, Mormyridae, Characidae and Cyprinodontidae) or not sufficiently specialized (e.g. Centropomidae) for this sort of anatomical change to be effective or even possible.

None will deny that, anatomically, these other families have undergone considerable evolution, in some cases more profound than that seen in the Cichlidae as a whole (the characids and mormyrids are spectacular examples). But this differentiation is of longer standing than that of the lake Cichlidae (witness the wide geographical distribution of these other species over much of Africa). Also this past differentiation has resulted in specializations that are inhibiting in terms of multiple adaptive radiation. The specialized dentition and jaws of the characids, the jaws of mormyrids, the edentulous but kinetically specialized jaws of the cyprinids and the specialized jaws of the cyprinodonts are all examples of a restriction in adaptive potential. Other inhibiting factors (especially breeding habits, see p. 12) were certainly operative in further restricting the evolution of these fishes in a lake environment.

In contrast, the syncranial characteristics of the Centropomidae (e.g. *Lates* and *Luciolates* species), although of a percoid grade, are in general far less specialized than those of a generalized *Haplochromis* species ; the jaw mechanism (see Alexander, 1967) is a particular case in point. Further, the syncranial organization in centropomids is a less suitable basis for the type of relative growth changes that have been so successful in the *Haplochromis* morphotype. Neither is the multiserial, microdont type of oral dentition in centropomids such an effective starting point for dental evolution as are the larger and fewer teeth of a haplochromine fish, and nor are the smaller, medially separated lower pharyngeal bones of a *Lates* such a suitable basis for the elaboration of a pharyngeal mill.

SPECIATION

All that has been discussed so far is proximately the product of one natural process, speciation. A great deal has been written about the background to, and mechanism of, speciation in African cichlid fishes. The subject is well reviewed by Fryer & Iles (1972), to whom reference should be made, especially for their detailed consideration of speciation in the cichlids of Lake Malawi.

Speciation involves two major but closely interrelated processes, genotype reorganization and the development of inherent barriers to reproduction with the mother species and related taxa. Genotypic reorganization requires that the population undergoing speciation should be isolated from genetic interchange with the stem species. It is this aspect of the problem that has received most attention in African cichlids (see below).

Little direct information is available on the evolution and nature of the barriers to interspecific crossing in these fishes. Whatever these barriers are, their effectiveness in the Lake Victoria *Haplochromis* species flock is very obvious. Despite the lack of definite and specific breeding seasons, and despite the close physical proximity of breeding sites occupied simultaneously by several species, there is virtually no evidence of interspecific hybridization (see p. 55). The effectiveness of these barriers is all the more remarkable when it is recalled that this flock of over 150 species has evolved, within little more than three-quarters of a million years, from one or at most a few closely related species. Clearly some, in genetical terms, simple mechanism is involved.

The breeding behaviour of several cichlids (including a few *Haplochromis* species, but unfortunately none from Lake Victoria) is now well known (Baerends & Baerends van Roon, 1950; Wickler, 1962a and b, 1963, 1966). Compared to that in many fishes it is complex, and in this complexity may lie a pointer to the 'ease' with which barriers to interspecific crosses are evolved. In other words, a slight deviation from an established pattern could provide an effective barrier to successful courtship and mating. The deviation might involve either behavioural patterns in courtship, or a specific recognition signal like male coloration (see p. 52). That there is no repetition of male coloration among the *Haplochromis* species of Lake Victoria points to the probable importance of this character as a specific recognition signal. Unfortunately we have no information on the genetical basis of coloration in these fishes nor on the mechanism whereby a female recognizes the male breeding colours of her species. Since young fishes of both sexes have a similar female-type coloration, and certainly are non-shoaling at an age when male breeding colours develop, it is unlikely that any element of imprinting is involved. Presumably the ability to recognize male conspecificity is inherent in females.

That differences in male breeding colours can evolve rapidly is well demonstrated by the endemic *Haplochromis* species of Lake Nabugabo (Greenwood, 1965b). This small lake was isolated from Lake Victoria by a sand spit formed some 3500 years B.P. Of the six *Haplochromis* species now occurring in the lake, five are endemic, but obviously related each to a species still living in Lake Victoria. Anatomically, the Nabugabo species are little different from their Victoria sister species (Greenwood, *op. cit.*). But, the differences in male coloration are very trenchant.

Since the Nabugabo species were derived from already fully differentiated Lake Victoria taxa isolated in the newly formed lake, there can be no grounds for thinking that the colour differences evolved in response to strong selection favouring characters that would prevent interspecific hybridization. In this respect the colour differences would seem merely to be one product of the genic reorganization undergone by the isolates. The significant feature, however, is the appearance of this character change in all but one of the isolated taxa, thus suggesting that, whatever its genetic basis, it is likely to occur whenever a genotype is reorganized.

If the situation in Lake Nabugabo is a typical consequence of isolating segments of a *Haplochromis* species, and if male coloration is of prime importance as a recognition signal (and there is little evidence to negate these suppositions) then we have a clear-cut example of the rapidity with which a *Haplochromis* species can originate.

Indeed, it seems that in these fishes speciation is an almost inevitable consequence of isolation, at least in a lacustrine environment.

This correlation of speciation and a lacustrine environment may have a dual basis. A lacustrine environment undoubtedly provides a greater number of spatially distinct, often repeated, niches than does a fluviatile one. Habitat preference could then lead to the isolation of populations, a situation which Fryer (1959 ; also Fryer & Iles, 1972) considers of first importance in the speciation of Lake Malawi cichlids. Secondly, a lake in its genesis is liable, through climatic or tectonic changes, to be broken up into smaller water bodies, or to give rise to peripheral isolates (like Lake Nabugabo). It is this type of background that I consider to have been involved in the speciation of the Lake Victoria flock (Greenwood, 1965a).

Much of the literature on evolution and speciation in Lake Victoria *Haplochromis* is concerned with the nature and origin of these primary isolating factors (see Regan, 1922 ; Worthington, 1937, 1954 ; Mayr, 1942, 1947 ; Brooks, 1950 ; Greenwood, 1951, 1959c, 1965a ; Hubbs, 1961 ; Kendall, 1969 ; Temple, 1969 ; and review in Fryer & Iles, 1972).

Apart from Mayr (1947), who then argued for the flock's origin by multiple colonization, the principal contention has been whether primary isolation was through habitudinal factors or by physical isolation in distinct water bodies. Mayr (1963) later acknowledged the unlikelihood of multiple colonization from rivers as a major factor in the evolution of these lake faunas (not specifically mentioning Lake Victoria amongst the examples he cites ; Grant [1963] still cites the earlier Mayr reference when dealing with the evolution of species swarms). Instead, Mayr (*op. cit.*) favoured extrinsic barriers within the lake as the most important factor in the isolation of populations within the lakes.

With the exception of Mayr (1963), Hubbs (1961) and Regan (1922), all other authors have agreed that some sort of gross physical isolation was the principal factor leading to speciation in these fishes. Hubbs (*op. cit.*) argues for ecological and temporal segregation, and Regan (*op. cit.*) for speciation and adaptive radiation through habitudinal segregation (*sensu* Gulick, 1905). It must be remembered that in Regan's time little was known of the lake's history, and that Regan's remarks on the evolution of Lake Victoria cichlids were also a reflection of his dissatisfaction with '. . . the modern view of evolution by accidental mutation'.

The role of habitudinal segregation in the speciation of Lake Victoria *Haplochromis* is difficult to establish. Few of the contemporary species show such close association with a particular habitat as do several of the Lake Malawi cichlids (Fryer & Iles, 1972), and none appears to be specialized for utilizing a food source that is itself restricted to a narrow habitat. It is, therefore, difficult to envisage any ecological barrier that might be sufficiently absolute to prevent some genic exchange between populations. In the present-day lake the major habitats themselves are not so clearly delimited as seems to be the case in Lakes Malawi and Tanganyika. Finally, and again unlike the two other lakes (see Fryer & Iles, 1972), there is no indication of any intralacustrine differentiation of populations ; thus, the implication is of genetical continuity of species' populations throughout the lake. Until fairly recently (see summaries in Kendall, 1969 ; also Doornkamp & Temple, 1966) thinking on the physiographical background to *Haplochromis* evolution in Lake Victoria was dominated by Wayland's (1934) Pluvial hypothesis. That is, the developing lake expanded and contracted (or was dried out completely) as a result of differing rainfall régimes during the Pleistocene. This theory of lake fractioning was first used by Brooks (1950) in his arguments to show that the Lake Victoria species flock could have evolved by means of the usually accepted process of allopatric speciation, rather than through sympatric (i.e. intralacustrine) means.

Wayland's theories were also basic to my first interpretation of speciation and adaptive radiation in the flock (Greenwood, 1951). In that paper I attempted to correlate what seemed to be the three structurally and ecologically definable major groups of endemic species with the three major Pluvial and Interpluvial periods influencing the lake's development. With increasing knowledge of the species' ecology and their phyletic relationships, and with new information on the lake's history, the conclusions I reached in that paper are no longer tenable. The inter-relationships of the fishes are far more complex, both ecologically and phyletically than I realized then, so that such a direct and simple correlation between lake development, adaptive radiation and phylogenesis cannot be achieved. Indeed, it seems that the three adaptive groups (Greenwood, *op. cit.*) are not only polyphyletic (i.e. are grades *sensu* Huxley, 1958) but that each evolved simultaneously. That is to say, there was concurrent speciation and differentiation of the trends occurring within the different phyletic lineages contributing to a grade.

The spatial isolation needed to separate the developing taxa now appears to be a correlate of both the way in which Lake Victoria originated and of later, tectonically induced, changes in the lake basin and its outflows (see p. 10 above ; also Greenwood, 1965a). Climatic changes, it seems, had little effect on the lake's early history and thus on the primary evolution of the *Haplochromis* species flock (Doornkamp & Temple, 1966 ; Temple, 1969 ; Kendall, 1969). Later, however, climatic effects could have played some part, especially in the isolation of small Nabugabo-like peripheral lakes (see Kendall, *op. cit.*).

A brief outline of Lake Victoria's origin (essentially from the results of river reversals and backponding) has already been given (see p. 10 ; also Greenwood, 1965a). On this basis Fryer & Iles (1972) constructed a detailed model of the way in which topographical events during the early period could have provided a variety of smaller and larger lakes, at times partially or wholly interconnected, at other times partially or wholly isolated from one another. The culmination of this history would, of course, be the union of separate lakes to form a single lake. But, even at this period there were lake-level changes leading to the isolation and reincorporation of small peripheral water-bodies (Kendall, 1969).

Initial differentiation of the main phyletic lines, and the development of trophic radiations within these lines must have taken place among the *Haplochromis* species inhabiting the shallow lakes first formed as the rivers were reversed and ponded back. Potentially, each isolate could be the cradle of a species, provided that sufficient time elapsed before the isolation was broken. That this time period might be a short one is demonstrated by the evolution of five species in Lake Nabugabo during a maximum time lapse of *ca* 3500 years (Greenwood, 1965a ; also p. 112 above).

At first the *Haplochromis* species would be in competition only with various non-cichlid species. By analogy with the non-cichlids of present-day rivers (Greenwood, 1966), these would include a number of specialized feeders (e.g. insectivorous Mormyridae and epilithic grazing cyprinids like *Labeo*) as well as more generalized feeders like the *Barbus* species, and the piscivorous catfishes (*Clarias* and *Bagrus* species). One can only presume from the trophically multiradiate *Haplochromis* species of the present lake that strong selective forces were operating in favour of specialized feeding habits. It is not possible to tell at what stage in the development of the flock these forces were most intense. Possibly this happened, not at first (when a generalized feeder might fare better against specialized competitors), but later when the number of *Haplochromis* species had increased. That is, after at least some of the original water-bodies with their contained *Haplochromis* species were interconnected. Thus, the competition would be between species of *Haplochromis*, a situation rendered more likely if the species tended to occupy habitats not especially favoured by the non-cichlid species (as is the situation in the present lake). It is even possible that the *Haplochromis* species occupied the more lacustrine areas of the developing lake, the non-cichlids occupying those regions which were environmentally more nearly fluviatile.

Once a few *Haplochromis* species had evolved, further fractioning of the growing lake would lead to an accelerated output of species. (Five species isolated in each of seven cut-off portions of the lake could give rise to 35 new species.)

As the number of species increased so it is reasonable to assume that selection pressures would rise in the competition for food and other niches, and with this a concomitant development of anatomical and physiological specializations.

The continuous spawning in *Haplochromis* populations (see p. 51) could be an important contributory factor in the development of such adaptations because it would expose to selective action a greater genetical turnover than would be the case in seasonally breeding species; advantageous gene combinations would also be spread more rapidly through a population. No data are available on the generation time for any Lake Victoria species, but for closely related species in Lake George, bred in aquaria, the figure is probably about two generations per year (Dr C. D. N. Barel, Leiden; personal communication).

The evolutionary literature is replete with references to the genetical effects of isolating small populations (for reviews see Mayr, 1963 and Grant, 1963). Although there are adverse effects, in terms of genic loss and also from genetically ill-adapted founder populations, the overall results of such population fractioning would seem to accord well with the situation found in present-day Lake Victoria. Possibly some of the adverse effects associated with small isolates might be mitigated by the speed with which *Haplochromis* can speciate, and the relatively short developmental period for Lake Victoria.

A noticeable feature of any adaptive trend within a *Haplochromis* lineage from this lake is the persistence of species showing intermediate stages in the evolution of a particular specialization (see pp. 57–93). This peculiarity would seem to suggest that during the evolution of a trend, selection pressure acting on sister derivatives was not at the same intensity. The size of the different water-bodies, the various combinations of species isolated together, and the probability of there being, at least for part of the lake's history, a large main body of water in which environmental conditions were more stable and selection less intense, are all possible factors allowing the persistence of graded specializations.

The existence in any one lineage (and at any adaptive grade in that lineage) of several closely related species distinguished by male coloration and slight morphometric differences is probably a reflection of their origin from populations of the same mother species isolated in different water-bodies. This mode of speciation would contradict Hennig's (1966) postulate that at a given point in time a stem species gives rise, dichotomously, to no more than two daughter species (one of which is the original stem species less that part of its genome incorporated in the new species). However, Brundin (1972) allows that an ancestral species may evolve into '. . . two or several daughter species', a view with which I would agree from my experience with the Lake Victoria *Haplochromis*. Later in the same paper Brundin expresses the view that simultaneous multiple splitting is a situation that cannot be conclusively demonstrated. Within the limits imposed by techniques available for identifying species and their relationships, this is undoubtedly true. But the Lake Victoria *Haplochromis* species do appear to be as good a *prima facie* case as can be found for simultaneous multiple splitting.

The development of the Lake Victoria *Haplochromis* flock cannot be construed as strictly intralacustrine for the majority of species and ecologically defined species groups. In this respect it differs from the cichlid flock of Lake Malawi, at least as far as the evolution of that flock is understood. Fryer & Iles (1972) argue cogently in favour of true intralacustrine speciation by habitat restriction and thus microgeographic isolation in Lake Malawi. However, I do not find all their arguments equally convincing, particularly since they are based essentially on a detailed study of only certain elements in the cichlid fauna ; the strength of habitat restriction is unknown for most of the *Haplochromis* species in Lake Malawi.

To me the weakest point in Fryer & Iles' argument lies in their failure to explain how the early invaders of the lake, species relatively unspecialized ecologically, were confined to and isolated in particular habitats. Isolated, that is, with sufficient effectiveness to prevent interpopulation gene flow. Even assuming that this did happen or that some type of stasipatric evolution (Key, 1968) occurred, and that the newly derived species became more closely tied to their habitats (as Fryer & Iles postulate) how, ultimately, did these stenotopic taxa manage to achieve lake-wide distribution? Undoubtedly some of the means suggested by Fryer & Iles (fluctuating lake levels in particular, *op. cit.* : 524) were operative. But there still seems to be a fundamental contradiction in an argument which invokes greater stenotopy in relatively unspecialized ancestors than in their observably stenotopic and highly specialized descendants (for example the several species that show interpopulational differences [Fryer & Iles, *op. cit.*] within Lake Malawi). Finally, I do not think that sufficient consideration has been given to conditions prevailing when Lake Malawi was first forming. Admittedly its present topography is very different from that of Lake Victoria, but was it necessarily a deep trench from its very inception? Could it not have passed through a period of shallow, isolated lakes like those at the start of Lake Victoria?

Despite these reservations, however, there is sufficient evidence both historical and contemporary (see Fryer & Iles, 1972) indicating that the cichlid flocks of Lakes Malawi and Victoria did have somewhat different evolutionary histories.

Mention was made above of the virtual absence of evidence for truly intralacustrine speciation in Lake Victoria. A possible exception may be provided by the several species confined to deeper waters (p. 49). Considering the physiographical history of the Lake, deepwater habitats were unlikely to have become available until the basin settled into more or less its present form. If that is so, then geographical barriers cannot be invoked to account for isolation of the evolving deepwater species. The alternative explanation would be an invasion of deep waters by individuals or populations of species occupying inshore habitats. Speciation initiated in this way would require that the migrant populations be able to establish breeding sites in their new habitats, and that no gene flow occurred between the in- and offshore communities. There is little evidence from contemporary species to support this hypothesis, and neither does it gain much support from what is known of the interrelationships of the deepwater species (see pp. 67–68). The presumed phylogeny of the '*tridens*' group of crustacean-eating species suggests an evolutionary history similar to that of any inshore lineage. That is, multiple simultaneous (or near simultaneous) speciation from a common stem. One can therefore only assume that the now deep living species evolved along with species from other habitats but, unlike the latter species, were able to colonize deep habitats when these became available. This question, like many others, should be left open and reconsidered when more is known about the deepwater species, particularly their range of depth tolerance in shallower regions of the offshore lake.

There has been a long-standing argument on the effects predatory species have had on the processes of speciation in African lakes. The debate was sparked off by Worthington (1937) who suggested that the presence of the large, piscivorous predators *Hydrocynus* (Characidae) and *Lates* (Centropomidae) in a lake inhibited speciation among the cichlids. His ideas derive from the fact that, excepting the Lake Tanganyika cichlid complex, cichlid species flocks only occur in lakes where *Lates* and *Hydrocynus* are absent. Lake Victoria in Worthington's view is a prime example of speciation unhampered by predators. Worthington, however, overlooked (or more correctly was unaware of) the predatory effects exerted by the numerous piscivorous *Haplochromis* species in the lake (see above p. 31) and those of the several non-cichlid predators also present (Greenwood, 1959c).

Palaeontological evidence (Greenwood, 1959d) shows that the *Haplochromis* species flocks of Lakes Edward and George (another key lake system in Worthington's hypothesis) evolved in the presence of *Lates* and *Hydrocynus*.

The physiographical history of Lakes Albert and Rudolf (lakes with rudimentary *Haplochromis* flocks, but with *Lates* and *Hydrocynus*) could just as well be responsible for the lack of speciation (see summary of the evidence in Fryer & Iles, 1972).

Indeed, the whole issue of predator—speciation interaction is ably and critically summarized by Fryer & Iles (*op. cit.*) and need not be discussed further here. In brief, there is no evidence supporting Worthington's original hypothesis (Worthington, 1937, 1954). On the contrary, the bulk of evidence suggests that predators are to be considered, if anything, factors promoting rather than inhibiting speciation (Greenwood, 1959c).

Connell & Orias (1964) have proposed an interesting ecological model to account for species diversity, especially in a tropical environment. The authors discount various commonly accepted ideas to explain this diversity, such as the number of available niches, rigorousness of the environment and the concept of 'biological immaturity' in temperature regions *versus* an equilibrium state in the tropics.

Instead they propose a model based essentially on the distribution of energy flow in physically stable environments contrasted with unstable environments. The tropics (stable) allow a greater proportion of energy to be used for net productivity (i.e. growth and reproduction) rather than for regulatory activities to counter environmental changes. From this, they argue, there will be larger populations, hence more genetic variation within populations, and thus better opportunities for successful speciation. Increased numbers of species will in turn provide a positive feedback by increasing environmental stability through, among other factors, increased recycling rates for nutrients. This feedback, Connell & Orias suggest, would be most marked in the early stages of evolution and would of course further the process of speciation. Ultimately, it is hypothesized, there would develop 'overspecialization' amongst the species produced, resulting in smaller populations and hence a negative feedback into the system.

The Connell–Orias model is based, implicitly, on homeothermic and terrestrial animals. Yet, there are interesting parallels with the situation, both historically and contemporaneously, in the Lake Victoria species flock. On the whole these fishes would seem to conform closely with the model. If increasing species diversity and number encourages further speciation and environmental stability, the rapidity with which *Haplochromis* speciates may contribute to the ecological success of these fishes and also account in part for their dominance in the lacustrine environment. Possibly in some taxa (e.g. the monotypic genera and other anatomically specialized species) we are seeing the 'overspecialization' that Connell & Orias consider as contributing to the negative feedback of the system.

The evolution of the *Haplochromis* species flock in Lakes Edward and George is usually linked with that of Lake Victoria (Trewavas, 1933 ; Greenwood, 1951, 1959d ; Fryer & Iles, 1972). There are two reasons for this. First, there is the great similarity existing between the species of the two lake systems (Lakes Edward and George are interconnected and can be considered one lake for the purpose of this discussion) ; this includes the sharing of certain *Haplochromis* and related species (Greenwood, 1973a). Second, the existence of river systems connecting the two lakes, and along which some transfer of fishes might have occurred in the past ; indeed, some geological evidence even suggests almost interlacustrine connections (Temple, 1969). These rivers now flow in two directions, eastward into Victoria and westward into Lake Edward–George. The watershed is a low swamp divide (Doornkamp & Temple, 1966). Today these swamps are impenetrable to all but air-breathing species, but they could have been less effective species filters in the past.

A recent revision of Lake George fishes (Greenwood, 1973a) shows that, although the *Haplochromis* species are closely related to those of Lake Victoria, there are fewer shared species than was once thought (Trewavas, 1933). The shared species are non-endemic taxa (i.e. *Haplochromis nubilus*, *Astatoreochromis alluaudi* and *Hemihaplochromis multicolor*). Recent geological studies (especially those of Doornkamp & Temple, 1966) show that although Lake Edward–George was filled from the same river systems as Lake Victoria, the Edward basin was established before that of Lake Victoria. Also, the Edward basin has been effectively isolated from that of Victoria for all of its existence.

The common fluviatile ancestry of the two lakes means, of course, that their species flocks also had, in large part, a common ancestory. The implications of this are manifold and include the possibility that *Lates* and *Hydrocynus* (known as fossils from Lake Edward, but no longer extant there ; Greenwood, 1959d) might also have been components of the early Lake Victoria biota (see p. 10 above). Unfortunately the Quaternary fossil record for Lake Victoria is virtually non-existent (Bishop, 1969) so this possibility cannot be checked.

A common ancestry for the Victoria and Edward *Haplochromis* species does allow one to make a meaningful comparison of the two species flocks (see Greenwood, 1973a). The pattern of adaptive radiation is extremely similar, with species showing all the trophic specializations found in the Lake Victoria flock (and, in addition, the evolution of a pelagic zooplankton eater). There are fewer species in Lake Edward–George (see Table I). The situation in Lake Edward, like that in Lake Nabugabo (see p. 112), could be looked upon as a live model of a phase in the evolution of the Lake Victoria flock. Lake Edward would then represent one of the larger preamalgam water-bodies (in which both speciation and adaptive radiation had taken place) and Lake Nabugabo a much smaller peripheral water-body in which differentiation had proceeded only to the level of speciation. That is, representative of the earlier and the later phases in the development of Lake Victoria.

No precise speciation rate can be determined for the *Haplochromis* of Lake Victoria. Dating of the lake's inception cannot be fixed more accurately than the mid-Pleistocene (Doornkamp & Temple, 1966 ; Temple, 1969), i.e. about 750 000 years B.P. Thus all one can say is that between 150 and 170 species have evolved in about three-quarter million years. Obviously the actual rate of transformation from one species to another must have been much faster. The figure derived from Lake Nabugabo (see Greenwood, 1965b) would indicate that a species can evolve in less than 3500 years, but even here we are still far from knowing the time taken to effect genetical isolation between mother and daughter species. If this figure is taken simply as an order of magnitude, and if due allowance is made for a tropical environment and the relatively short generation time in *Haplochromis*, speciation rate is about one and a half times to twice that in the few other genera of fishes (all non-cichlids) for which comparable data are available (Myers, 1960 ; Hubbs & Raney, 1946).

CONCLUSION

There has been a tendency to consider the African cichlid species flocks as a somewhat unique evolutionary phenomenon (see p. 3). What justification is there for this attitude, and what contribution do these flocks make to understanding the broader picture of speciation and evolution among freshwater fishes, and in general?

It is usual to compare the cichlids to the Galapagos Island Finches (Darwin's Finches) and to the Hawaiian Honeycreepers. There are indeed many broad similarities between these bird and fish, island and lake, species flocks. In both situations the main evolutionary trends have been towards trophic specialization (with concomitant anatomical specializations; see Lack, 1947; Amadon, 1950). The comparison between the lakes and islands as vacant, geographically isolated, multi-niche environments is also a valid one. This latter point is particularly evident when it is recalled that, during its early history, Lake Victoria was a group of lakes and thus geographically akin to the background of evolution in the two bird groups, more so than would be a single lake. Fryer & Iles (1972) argue, however, that marked habitat differentiation in such lakes as Malawi and Tanganyika makes them in effect multiple and not single water-bodies. Be that as it may, the end point of speciation in these lakes is basically little different from that in Lake Victoria.

The range of anatomical and ecological diversity found amongst the cichlid fishes far surpasses that of the Finches and Honeycreepers. This I believe is explicable on an anatomical basis alone. The fish syncranium is a better starting point for trophically orientated morphological modification than is a bird skull. Especially is this so for a cichlid fish where the jaws, teeth and pharyngeal apparatus have reached a level of differentiation that gives the system a high degree of morpho-potential (see pp. 103–111 above). Other factors, both inherent and environmental, can be adduced to explain the greater differentiation and speciation of the Great Lake cichlid fishes (see p. 111).

It is believed that one factor encouraging adaptive radiation in island invading birds was the lack of competition in their new habitat, especially when compared to conditions on the mainland. Conditions for fluviatile cichlid species invading a developing lake might be different. Unlike the lone bird species colonizing an island, the ancestral river *Haplochromis* species would enter the new environment in the company of several different kinds of fishes. These would include, particularly, members of various non-cichlid families such as the Cyprinidae, Mormyridae and the catfishes.

In this assemblage, so it is argued (Fryer & Iles, 1972), there is potential competition for the one or two founder cichlid species. However, since the colonizers had already established some kind of equilibrium in the river environment, and since the developing lake would differ little from a river, there might, in fact, be little competition at first. That is to say, there would be pre-emption of certain habitats from the outset by the differently adapted invaders. Only after the cichlids had speciated would competition become a major factor in their evolution. In this respect I believe that there is really little difference between the island birds and the lake fishes. However, I would agree with Fryer & Iles when they point out that one major difference lies in the fact that the cichlids, unlike the birds, evolved in the presence of predators; but the consequences of that interaction are debatable; see above, p. 117.

All in all, the cichlid species flocks, despite the similarities in evolutionary background, are more complex and probably more successful than are the species flocks of Galapagos Geospizidae and Hawaiian Drepaniidae.

A similar conclusion is reached when the cichlid flocks are compared with the few other examples of lacustrine species swarms among fishes (including those of the non-cichlids occurring sympatrically with the cichlids). The best studied of these are the cyprinid flock of Lake Lanao in the Philippines (18 endemic species ; see Myers, 1960), the 14 endemic species of the cyprinodont subfamily Orestiinae in Lake Titicaca (Tchernavin, 1944 ; Brooks, 1950) and the 18 endemic cottoid species of Lake Baikal (Berg, 1925, 1928 ; Brooks, 1950 ; Kozhov, 1963). Lesser flocks are discussed by Myers (1960) and Mayr (1963) ; all are completely overshadowed by the cichlid flocks, as indeed are all other vertebrate and most invertebrate species swarms. Exceptional among the latter are the gammarid Crustacea of Lake Baikal (Brooks, 1950 ; Kozhov, 1963) and, particularly, the drosophilid flies of Hawaii. The similarities between the flies and the African cichlids are remarkably close, involving localized species swarms of up to 160, mostly endemic, species, and much morphological diversity associated with trophic specialization (see discussion in Fryer & Iles, 1972).

Thus, an answer to the first question posed at the beginning of this section seems to be largely in the negative. The cichlid species flocks are not a unique evolutionary phenomenon, nor has their evolution involved organic or physical processes unique to these animals or the environment in which they occur. But, the cichlids are outstanding examples of this particularly concentrated and tachytelic type of evolution, be they compared with vertebrate or invertebrate animals. The reasons for this singularity seem to lie in the fortuitous combination of environmental circumstances, the great morpho-potential of most cichlid taxa, and the biology (especially the reproductive biology) of these taxa.

Turning now to the question of what light the species flocks throw on evolutionary processes in general, there is one concept that in part reflects back onto the presumed peculiarities of lacustrine species flocks, and in part has broader application.

Myers (1960) drew attention to a peculiar feature of many endemic lake fish faunas, of which the cichlid flocks (especially those of Lakes Malawi and Tanganyika) provide outstanding examples. This feature, which Myers called 'supralimital specialization', involves the evolution of characters in taxa, of circumscribed distribution, that transcend the morphological limits of a related taxon's systematic category elsewhere in its range. Myers is particularly concerned with what might be termed 'suprafamilial' character transgression, but it is clear that he applies the term to situations transcending generic limits as well.

Philosophically, one may jib at the circularity of this concept, but its observational basis is undeniable. In Lake Victoria the monotypic genera *Platytaeniodus* and *Hoplotilapia* (and even *Macropleurodus*) would qualify as supralimital taxa, and there are even better examples among the species of Lakes Tanganyika and Malawi (Poll, 1956 ; Fryer & Iles, 1972).

According to Myers (*op. cit.* : 332) supralimitally specialized species are '. . . often capable of becoming the founders of new genera, families or perhaps even higher categories, at new adaptive levels. They have unquestionably done so in the older lake fish faunas. . . .' The importance of this statement would seem to lie in the implication that the development of supralimital specializations lead to evolutionary success because their possessors are able to enter new adaptive levels. It is this aspect of Myer's concept that requires careful consideration.

In Lake Victoria there are several examples of supralimital specialization. At the species level there are *Haplochromis xenognathus* (see p. 70), *H. welcommei* (p. 69) and *H. aelocephalus* (p. 61 ; also Greenwood, 1959b) ; at the generic level, *Macropleurodus bicolor, Platytaeniodus degeni* and *Hoplotilapia retrodens* (see pp. 99–103). As might be expected (and as is implied in Myer's formulation of the supralimital phenomenon), these morphologically distinctive taxa also show definite ecological specializations as well, namely in these examples, trophic specialization. Each example represents an evolutionary break-through into a new adaptational level. But equally distinctive breaks have been made by many other Victoria *Haplochromis* species which do not exhibit morphological changes that could be termed supralimital. It would, for example, be difficult to consider the mollusc-crushing species (p. 72), the algal grazers (p. 63) or the piscivores (p. 80) as transcending generic limits unless one took a narrowly typological view of the flock's taxonomy. (Ironically, the type species for the genus *Haplochromis* is *H. obliquidens*, a species with highly specialized teeth ; see p. 65).

In other words, the Lake Victoria *Haplochromis* species have succeeded in occupying several adaptational levels (levels that are new relative to that of the ancestral fluviatile species) without producing supralimital specializations. Expressed in another way, supralimital specialization is not a necessary prerequisite for evolutionary success nor is it necessarily a common way of achieving it.

Myers (*op. cit.* : 331) also implies that the full success of supralimital specializations can only be exploited outside their place of origin : 'If they (the specialized species) could get out of their lakes and use their supralimital specializations in other lakes or streams, . . . many existing lake fishes could easily become the founders of large and flourishing new groups at new adaptational levels'. I submit that the lake cichlids have done this, and *in situ* ; also, that it is unnecessary to link such success with altered taxonomic status (i.e. marked morphological change) for the animal achieving it. What strikes me as more important than extreme morphological change *per se* is the evolutionary opportunity offered by lakes (or a similar type of evolutionary background) in the development of ecological specialization. In this sense I would endorse Myer's ideas on the probable importance of lakes (or near lacustrine conditions) in the evolution of fish faunas now no longer lacustrine but, as it were, secondarily fluviatile. For example, the characid fishes of tropical America (cited by Myers, 1960) and probably several of the endemic species groups in the Congo river (see Roberts, 1972 : 128).

The reasons for the existence of 'supralimital' morphotypes in species flocks remains to be explained, especially since, as in Lake Victoria, apparent adaptational success can be achieved without extreme anatomical modification. With only the information currently available it is, of course, impossible to really evaluate this success, and each example of a 'successful' species can only be considered in the context of its particular environment. Thus it is possible that during the evolution of various biotopes in Lakes Tanganyika and Malawi more intense selection has resulted in the greater morphological differences seen in the species of those lakes than in those of Lake Victoria. In other words, a specialization has to be more specialized if it is to succeed in the former lakes. Once again, a dearth of ecological information for species of Lakes Malawi and Tanganyika hampers speculation.

Myer's views (1960) on the importance of tachytelic supralimital specialization is in direct conflict with an hypothesis put forward by Briggs (1966). According to Briggs there are two, simultaneously occurring kinds of evolutionary change. One (the so-called successful type) is slow and likely to produce species with a potential phyletic future, the other (so-called unsuccessful) is rapid and unproductive of species with phyletic potential.

Briggs (*op. cit.*) associates the centres of unsuccessful evolution with peripheral areas of a species' distribution, and with evolutionary traps (*sensu* Simpson, 1953), like lakes and islands. It cannot be denied that a lake is potentially a trap (both in the physical and evolutionary senses of that word). Equally, it cannot be overruled as a source of phyletically potential taxa if its fauna is later dispersed (see above; also Myers, 1960, and Roberts, 1972, for possible examples of this having happened). Again one faces the difficulty of assessing evolutionary success. At the present moment lake cichlids are a successful group in that environment, and I can find no *a priori* reason why some of these species could not be successful if they were to invade a major new system. That the fluviatile fishes are not generally outstanding in their diversity (but recall the Congo and Amazon; p. 4) is probably attributable to the absence, historically, of these physio-geographic circumstances that characterize lake formation and development.

Briggs (following Bates [1960]) cites, as a general evolutionary principle, the greater stability of communities composed of diverse types and numerous species (see also the ideas of Connell & Orias [1964] discussed on p. 118). From this, Briggs argued that stable ecosystems are the most important centres for successful evolution. The unstable beginnings of Lake Victoria seem to have had important evolutionary results, and it is very probable that the early environments (ecologically and physiographically) of Lakes Malawi and Tanganyika were also unstable. The end product in all three lakes, however, was a diverse multi-specific ecosystem of apparent stability. Furthermore, this has been achieved, even in the older lakes, at a rate which can only be described as tachytelic.

Thus, one might postulate, from the viewpoint of evolutionary potential, that situations akin to those of the African lakes are important because their faunas are able to reach a level of diversity, and therefore ultimate stability, in a short time.

Like the concept of supralimital specialization, that of Briggs' 'successful' and 'unsuccessful' evolutionary change seems as much semantic as biological, with the added difficulty of distinguishing between evolutionary success and phyletic longevity. Can one really compare, in terms of evolutionary success, the coelacanths and teleosts?

Without doubt factors used by Myers and Briggs to support their opposing arguments have played a part in the ebb and flow of phylogenesis. Perhaps the most realistic conclusion is that of Romer (1960), who agreed that explosive evolution (i.e. rapid diversification) is often followed by extinction, but that it is not an exceptional process nor, for most groups, an unusual one.

In this, as in so many other respects, the cichlid species flocks are evolutionary microcosms repeating on a small and appreciable scale the patterns and mechanisms of vertebrate evolution. It is interesting, in this context, to quote the last paragraph of the summary from Grant's book *The origin of adaptations* (1963): 'It follows that long continued evolutionary trends in some phyletic lines which undergo rapid and continual formation of new adaptive allelle combinations may be expected to follow a course of repeated speciational branchings. The phyletic line in such cases progresses through a steplike succession of divergences, in which each daughter species with its particular new adaptive allelle combination, branches off from a genetically different and more conservative ancestral population, and later gives rise in its turn to another new daughter species which diverges again in the same general direction.'

The aim of Grant's book was '. . . to set forth the causal theory of evolution as applied to diploid sexual organisms'. Yet, the paragraph quoted above could well describe the three-quarter million year old *Haplochromis* species flock in Lake Victoria.

ACKNOWLEDGEMENTS

Very many people have contributed in one way or another to this study. It is with pleasure that I offer them my thanks and deep appreciation for their help, intellectual or physical (and in the case of my colleagues in east Africa, often of both kinds).

I am particularly indebted to two people. First, to my former teacher Professor D. J. Nolte of the Witwatersrand University, South Africa, who started it all by stimulating my interest in problems of speciation and evolution. Next, Dr Ethelwynn Trewavas (then Curator of Fishes in this Museum), whose pioneering and wide-ranging researches into African cichlids were one of my early stimuli and have remained as a constant inspiration. Her kindness and knowledge were unstintingly given to me, first as a student and then over many years of close association as colleagues. I shall always be grateful to her for suggesting that I study the *Haplochromis* of Lake Victoria, and for the time and patience she expended on my introduction to these fishes (and to ichthyology in general).

My work in the field, as a member of the East African Fisheries Research Organization (1951–58), was strongly supported by its director Mr R. S. A. Beauchamp; I take this opportunity to thank him not only for this, but for his appreciation of the need to carry out basic taxonomic work on the freshwater fishes of east Africa. My best thanks go also to Dr R. H. Lowe-McConnell and Dr Philip S. Corbet, former colleagues at E.A.F.R.O., with whom I collaborated on many ecological aspects of *Haplochromis* biology.

More recently I have received a great deal of assistance from the current director of E.A.F.R.O. (now the East African Freshwater Fisheries Research Organization), Dr John Okedi, his staff (in particular Dr Michael Gee), and the director of the

F.A.O./U.N.D.P. team, Mr Peter Jackson and his staff, working with E.A.F.F.R.O· on the fisheries of Lake Victoria.

In the British Museum (Natural History) I am particularly indebted to Mr Gordon Howes on whose willing shoulders and into whose capable hands have fallen much of the drudgery in preparing this paper. Most of the osteological and anatomical figures are his work, and for this alone I am especially grateful.

To the many people not mentioned by name I can only repeat my thanks for their considerable help in so many different ways.

REFERENCES

ALEXANDER, R. McN. 1967. The functions and mechanisms of the protrusible upper jaws of some acanthopterygian fish. *J. Zool. Lond.* **151** : 43–64.

AMADON, D. 1950. The Hawaiian honeycreepers (Aves, Drepaniidae). *Bull. Am. Mus. nat. Hist.* **95** : 151–262.

BAERENDS, G. P. & BAERENDS VAN ROON, J. M. 1950. An introduction to the study of the ethology of cichlid fishes. *Behaviour Suppl.* **1** : 1–242.

BATES, M. 1960. Ecology and evolution. In *Evolution after Darwin*, S. Tax (ed.), **1** : 547–568. Chicago.

BEAUCHAMP, R. S. A. 1964. The Rift Valley lakes of Africa. *Verh. Int. ver. Limnol.* **15** : 91–99.

BERG, L. S. 1925. Die Fauna des Baikalsees und ihre Herkunft. *Arch. Hydrobiol., Suppl.* **4** : 479–526.

—— 1928. New data on the origin of the Baikalian fauna. (In Russian.) *Dokl. Acad. Sci. USSR, A*, **22** : 459–464.

BISHOP, W. W. 1969. Pleistocene stratigraphy in Uganda. *Mem. geol. Surv. Uganda*, no. 10 : 1–128.

BISHOP, W. W. & TRENDALL, A. F. 1967. Erosion-surfaces, tectonics and volcanic activity in Uganda. *Q. J. geol. Soc. Lond.* **122** : 385–420.

BOCK, W. J. 1959. Preadaptation and multiple evolutionary pathways. *Evolution*, **13** : 194–211.

BRIGGS, J. C. 1966. Zoogeography and evolution. *Evolution*, **20** : 282–289.

BROOKS, J. L. 1950. Speciation in ancient lakes. *Q. Rev. Biol.* **25** : 30–60 & 131–176.

BROWN, W. L. & WILSON, E. O. 1956. Character displacement. *Syst. Zool.* **5** : 49–64.

BRUNDIN, L. 1972. Evolution, causal biology and classification. *Zoologica Scr.*, **1**, 3–4 : 107–120.

CONNELL, J. H. & ORIAS, E. 1964. The ecological regulation of species diversity. *Am. Nat.* **98** : 399–414.

CORBET, P. S. 1960. Breeding sites of non-cichlid fishes in Lake Victoria. *Nature, Lond.* **187** : 616–617.

—— 1961. The food of non-cichlid fishes in the Lake Victoria basin, with remarks on their evolution and adaptation to lacustrine conditions. *Proc. zool. Soc. Lond.* **136** : 1–101.

DOORNKAMP, J. C. & TEMPLE, P. H. 1966. Surface, drainage and tectonic instability in part of southern Uganda. *Geogrl J.* **132** : 238–252.

ELTON, C. 1928. *Animal ecology*. London.

FISH, G. R. 1951. Digestion in *Tilapia esculenta*. *Nature, Lond.* **167** : 900.

—— 1955. The food of *Tilapia* in East Africa. *Uganda J.* **19** : 85–89.

—— 1957. A seiche movement and its effect on the hydrology of Lake Victoria. *Colon. Off. Fish. Publ. Lond.* **10** : 1–68.

FRYER, G. 1959. The trophic interrelationships and ecology of some littoral communities of Lake Nyasa with special reference to the fishes, and a discussion of the evolution of a group of rock-frequenting Cichlidae. *Proc. zool. Soc. Lond.* **132** : 153–281.

FRYER, G. 1961. Observations on the biology of the cichlid fish *Tilapia variabilis* Boulenger in the northern waters of Lake Victoria (East Africa). *Revue Zool. Bot. afr.* **64** : 1–33.

FRYER, G. & ILES, T. D. 1969. Alternative routes to evolutionary success among African cichlid fishes, as shown by the genus *Tilapia* and the species flocks of the great lakes. *Evolution*, **23** : 359–369.

—— 1972. *The cichlid fishes of the Great Lakes of Africa. Their biology and evolution.* Edinburgh.

GARROD, D. J. 1959. The growth of *Tilapia esculenta* Graham in Lake Victoria. *Hydrobiologia*, **12** : 268–298.

GEE, J. M. 1968. Trawling operations on Lake Victoria. *Rep. E. Afr. Freshwat. Fish. Res. Org.* **1967** : 55–63.

GRAHAM, M. 1929. *A report on the fishing survey of Lake Victoria*, 1927–1928, *and appendices.* Crown Agents, London.

GRANT, V. 1963. *The origin of adaptations.* New York.

GREENWOOD, P. H. 1951. Evolution of the African cichlid fishes : the *Haplochromis* species-flock in Lake Victoria. *Nature, Lond.* **167** : 19–20.

—— 1953. Feeding mechanism of the cichlid fish *Tilapia esculenta* Graham. *Nature, Lond.* **172** : 207–208.

—— 1954. On two species of cichlid fishes from the Malagarazi river (Tanganyika), with notes on the pharyngeal apophysis in species of the *Haplochromis* group. *Ann. Mag. nat. Hist.* (12) **7** : 401–414.

—— 1956a. The monotypic genera of cichlid fishes in Lake Victoria. *Bull. Br. Mus. nat. Hist.* (Zool.), **3** : 295–333.

—— 1956b. A revision of the Lake Victoria *Haplochromis* species (Pisces, Cichlidae), Part I. *Bull. Br. Mus. nat. Hist.* (Zool.), **4** : 223–244.

—— 1957. A revision of the Lake Victoria *Haplochromis* species (Pisces, Cichlidae), Part II. *Bull. Br. Mus. nat. Hist.* (Zool.), **5** : 73–97.

—— 1958a. A new genus and species of catfish (Pisces, Clariidae) from the deeper waters of Lake Victoria. *Ann. Mag. nat. Hist.* (13) **1** : 321–325.

—— 1958b. Reproduction in the east African lung-fish *Protopterus aethiopicus* Heckel. *Proc. zool. Soc. Lond.* **130** : 547–567.

—— 1959a. The monotypic genera of cichlid fishes in Lake Victoria, Part II. *Bull. Br. Mus. nat. Hist.* (Zool.), **5** : 163–177.

—— 1959b. A revision of the Lake Victoria *Haplochromis* species (Pisces, Cichlidae), Part III. *Bull. Br. Mus. nat. Hist.* (Zool.), **5** : 179–218.

—— 1959c. Evolution and speciation in the *Haplochromis* (Pisces, Cichlidae) of Lake Victoria. *15th Int. Congr. Zool.* **1958** : 147–150.

—— 1959d. Quaternary fish-fossils. *Explor. Parc. natn. Albert Miss. J. de Heinzelin de Braucourt*, **4** : 1–80.

—— 1960. A revision of the Lake Victoria *Haplochromis* species (Pisces, Cichlidae), Part IV. *Bull. Br. Mus. nat. Hist.* (Zool.), **6** : 227–281.

—— 1962. A revision of the Lake Victoria *Haplochromis* species (Pisces, Cichlidae), Part V. *Bull. Br. Mus. nat. Hist.* (Zool.), **9** : 139–214.

—— 1965a. Explosive speciation in African lakes. *Proc. R. Instn Gt. Br.* **40** : 256–269.

—— 1965b. The cichlid fishes of Lake Nabugabo, Uganda. *Bull. Br. Mus. nat. Hist.* (Zool.), **12** : 315–357.

—— 1965c. Environmental effects on the pharyngeal mill of a cichlid fish, *Astatoreochromis alluaudi*, and their taxonomic implications. *Proc. Linn. Soc. Lond.* **176** : 1-10.

—— 1966a. *The fishes of Uganda* (2nd edition). Uganda Society, Kampala.

—— 1966b. Two new species of *Haplochromis* (Pisces, Cichlidae) from Lake Victoria. *Ann. Mag. nat. Hist.* (13) **8** : 303–318.

—— 1967. A revision of the Lake Victoria *Haplochromis* species (Pisces, Cichlidae), Part VI. *Bull. Br. Mus. nat. Hist.* (Zool.), **15** : 29–119.

—— 1971. On the cichlid fish *Haplochromis wingatii* (Blgr.), and a new species from the Nile and Lake Albert. *Revue Zool. Bot. afr.* **84** : 344–365.

—— 1973a. A revision of the *Haplochromis* and related species (Pisces, Cichlidae) from Lake George, Uganda. *Bull. Br. Mus. nat. Hist.* (Zool.), **25** : 139–242.

—— 1973b. Morphology, endemism and speciation in African cichlid fishes. *Verh. dt. zool. Ges. Mainz,* **1973** : 115–124.

GREENWOOD, P. H. & GEE, J. M. 1969. A revision of the Lake Victoria *Haplochromis* species (Pisces, Cichlidae), Part VII. *Bull. Br. Mus. nat. Hist.* (Zool.), **18** : 1–65.

GULICK, J. T. 1905. Evolution, racial and habitudinal. *Publs Carnegie Instn.* no. 25 : 1–269.

HENNIG, W. 1966. *Phylogenetic systematics.* Urbana.

HUBBS, C. L. 1961. Isolating mechanisms in the speciation of fishes. In *Vertebrate speciation,* W. F. Blair (ed.) : 5–23. Austin.

HUBBS, C. L. & RANEY, E. C. 1946. Endemic fish fauna of Lake Waccamaw, North Carolina. *Misc. Publs. Mus. Zool. Univ. Mich.* **65** : 1–30.

HUXLEY, J. S. 1958. Evolutionary processes and taxonomy with special reference to grades. *Uppsala Univ. Årsskr.,* **1958** : 21–39.

KEAST, A. & WEBB, D. 1966. Mouth and body form relative to feeding ecology in the fish fauna of a small lake, Lake Opinicon, Ontario. *J. Fish Res. Bd Can.* **23** : 1845–1874.

KENDALL, R. L. 1969. An ecological history of the Lake Victoria basin. *Ecol. Monogr.* **39** : 121–176.

KEY, K. H. L. 1968. The concept of stasipatric speciation. *Syst. Zool.* **17** : 14–22.

KOZHOV, M. 1963. *Lake Baikal and its life.* The Hague.

LACK, D. 1947. *Darwin's finches.* Cambridge.

LIEM, K. F. 1970. The comparative anatomy of the Nandidae (Pisces : Teleostei). *Fieldiana Zool.* **56** : 1–166.

LOWE, R. H. 1952. Report on the *Tilapia* and other fish and fisheries of Lake Nyasa 1945–47. *Colon. Off. Fish. Publ. Lond.* **1** (2) : 1–126.

—— 1953. Notes on the ecology and evolution of Nyasa fishes of the genus *Tilapia*, with a description of *T. saka* Lowe. *Proc. zool. Soc. Lond.* **122** : 1035–1041.

LOWE-MCCONNELL, R. H. 1956. Observations on the biology of *Tilapia* (Pisces–Cichlidae) in Lake Victoria, East Africa. *E. Afr. Freshwat. Fish. Res. Org. Suppl. Publc.* (1) : 1–72.

—— 1969. Speciation in tropical freshwater fishes. *Biol. J. Linn. Soc.* **1** : 51–75.

MAYR, E. 1942. *Systematics and the origin of species.* New York.

—— 1947. Ecological factors in speciation. *Evolution,* **I** : 268–288.

—— 1963. *Animal species and evolution.* Harvard.

—— 1969. *Principles of systematic zoology.* McGraw-Hill, London.

MORIARTY, D. J. W. & MORIARTY, C. M. 1973. The assimilation of carbon from phytoplankton by two herbivorous fishes : *Tilapia nilotica* and *Haplochromis nigripinnis.* *J. Zool. Lond.* **171** : 41–55.

MYERS, G. S. 1960. The endemic fish fauna of Lake Lanao, and the evolution of higher taxonomic categories. *Evolution,* **14** : 323–333.

NEIL, E. H. 1964. An analysis of color changes and social behaviour of *Tilapia mossambica.* *Univ. Calif. Publs. Zool.* **75** : 1–58.

PLATE, L. 1913. *Selektionsprinzip und Probleme der Artbildung.* Leipzig and Berlin.

POLL, M. 1956. Poissons Cichlidae. *Résult. scient. Explor. hydrobiol. lac Tanganyika* (1946–1947), **3**, fasc. 5b : 1–619.

REGAN, C. T. 1920. The classification of the fishes of the family Cichlidae. I. The Tanganyikan genera. *Ann. Mag. nat. Hist.* (9) **5** : 33–53.

—— 1921. The cichlid fishes of Lake Nyasa. *Proc. zool. Soc. Lond.* **1921** : 675–727.

—— 1922. The cichlid fishes of Lake Victoria. *Proc. zool. Soc. Lond.* **1922** : 157–191.

RENSCH, B. 1933. Zoologische Systematik und Artbildungsproblem. *Verh. dt. zool. Ges. Köln,* **1933** : 19–83.

ROBERTS, T. R. 1972. Ecology of fishes in the Amazon and Congo basins. *Bull. Mus. comp. Zool. Harv.* **143** : 117–147.

ROMER, A. S. 1960. Explosive evolution. *Zool. Jb.* (*Syst.*), **88** : 79–90.
SIMPSON, G. G. 1953. *The major features of evolution.* New York.
TALLING, J. F. 1963. Origin of stratification in an African rift lake. *Limnol. Oceanogr.* **8** : 68–78.
TALLING, J. F. & TALLING, I. B. 1965. The chemical composition of African lake waters. *Int. Revue ges. Hydrobiol. Hydrogr.* **50** : 421–463.
TCHERNAVIN, V. V. 1944. A revision of the subfamily Orestiinae. *Proc. zool. Soc. Lond.* **114** : 140–233.
TEMPLE, P. H. 1969. Some biological implications of a revised geological history for Lake Victoria. *Biol. J. Linn. Soc.* **1** : 363–371.
TREWAVAS, E. 1933. Scientific results of the Cambridge expedition to the East African lakes, 1930–1. The cichlid fishes. *J. Linn. Soc.* (*Zool.*), **38** : 309–341.
—— 1935. A synopsis of the cichlid fishes of Lake Nyasa. *Ann. Mag. nat. Hist.* (10), **16** : 65–118.
—— 1938. Lake Albert fishes of the genus *Haplochromis*. *Ann. Mag. nat. Hist.* (11) **1** : 435–449.
—— 1949. The origin and evolution of the cichlid fishes of the great African lakes, with special reference to Lake Nyasa. *13th Int. Congr. Zool.* **1948** : 365–368.
TREWAVAS, E., GREEN, J. & CORBET, S. A. 1972. Ecological studies on crater lakes in West Cameroon. Fishes of Barombi Mbo. *J. Zool., Lond.* **167** : 41–95.
WAYLAND, E. J. 1934. Rifts, rivers, rains and early man in Uganda. *Jl R. anthrop. Inst.* **64** : 333–352.
WEATHERLEY, A. H. 1963. Notions of niche and competition among animals, with special reference to freshwater fish. *Nature, Lond.* **197** : 14–17.
WHITEHEAD, P. J. P. 1959. The anadromous fishes of Lake Victoria. *Revue Zool. Bot. afr.* **54** : 329–363.
WICKLER, W. 1962a. Zur Stammesgeschichte funktionell korrelierter Organ- und Verhaltensmerkmale : Ei-Attrapen und Maulbrüten bei afrikanischen Cichliden. *Z. Tierpsychol.* **19** : 129–164.
—— 1962b. 'Egg dummies' as natural releasers in mouth-breeding cichlids. *Nature, Lond.* **194** : 1092–1093.
—— 1963. Zur Klassifikation der Cichlidae, am Beispiel der Gattungen *Tropheus*, *Petrochromis*, *Haplochromis* und *Hemihaplochromis* n. gen. (Pisces, Perciformes). *Senckenberg. biol.* **44** : 83–96.
—— 1964. Das Problem der stammesgeschichtlichen Sackgassen. *Naturw. Medizin.* **1** : 16–29.
—— 1966. Sexualdimorphismus, Paarbildung und Versteckbrüten bei Cichliden (Pisces : Perciformes). *Zool. Jb. Syst.* **93** : 127–138.
—— 1968. *Mimicry in plants and animals.* London.
WORTHINGTON, E. B. 1937. On the evolution of fish in the great lakes of Africa. *Int. Revue ges. Hydrobiol. Hydrogr.* **35** : 304–317.
—— 1954. Speciation of fishes in African lakes. *Nature, Lond.* **173** : 1064–1067.
WYNNE-EDWARDS, V. C. 1962. *Animal dispersion in relation to social behaviour.* Edinburgh.

P. H. GREENWOOD, D.Sc.
Department of Zoology
BRITISH MUSEUM (NATURAL HISTORY)
CROMWELL ROAD
LONDON SW7 5BD

INDEX

N.B. Page numbers in *italics* refer to illustrations.